Handbook of

Power and Distribution Transformer Practices

Handbook of

Power and Distribution Transformer Practices

H Kailasaraman

Universities Press

UNIVERSITIES PRESS (INDIA) PRIVATE LIMITED

Registered Office
3-6-747/1/A & 3-6-754/1, Himayatnagar
Hyderabad 500 029, Telangana, India
email: info@universitiespress.com, www.universitiespress.com

Distributed by
Orient Blackswan Private Limited

Registered Office
3-6-752 Himayatnagar, Hyderabad 500 029, Telangana, India

Other Offices
Bengaluru / Bhopal / Chennai / Ernakulam / Guwahati / Hyderabad /
Jaipur / Kolkata / Lucknow / Mumbai / New Delhi / Noida / Patna

ISBN 978 81 7371 984 4

Typeset in Life BT at 10.6pt *by*
OSDATA, Hyderabad 500 029

Printed in India by
Graphica Printers and Binders
Hyderabad 500 013

Published by
Universities Press (India) Private Limited
3-6-747/1/A & 3-6-754/1, Himayatnagar
Hyderabad 500 029, Telangana, India

Contents

Foreword

Transmission and distribution are vital elements of the power sector besides generation. But for efficient transmission and distribution, the power generated cannot reach consumers without loss and waste. For the right quality and quantity of power, utilities and facilities depend on the efficiency and effectiveness of the transmission system.

Power transformers and autos play an important role in transmission networks. In this book titled *Handbook of Power and Distribution Transformer Practices*, H. Kailasaraman elucidates this role in an exemplary manner. His practical experience in Tamil Nadu Electricity Board, his thirst for theoretical knowledge in this field and his experience as an engineer in the training centre have all helped him in authoring this excellent guide for engineers in the transmission and distribution field. Apart from explaining the basic theories and the features of power transformers and their accessories, he has given practical tips as preventive and remedial measures for many field situations.

I believe that engineers working in various electricity boards and private sector utilities stand to gain a lot from this book. If the guidelines suggested by the author are adopted, interruptions can be avoided to a large extent and savings in power loss can be ensured. The useful lifetime of equipments can also be extended significantly and accidents avoided.

I hope all the intended readers will make good use of the book and that it will give an impetus to the efficient functioning of the transmission and distribution wing of the power sector.

S. Sundara Srinivasan, BE, FIE
Former Chief Engineer (Operation)
Tamil Nadu Electricity Board

Preface

Just as the bicycle has not undergone any significant change in its basic design since its invention, transformers too have not undergone any major changes except for the technological advancement brought about by the improved quality of materials. Transformers are widely handled by power engineers of power supply utilities and industries. This handbook of power and distribution transformers has resulted from my desire to share with practising engineers the knowledge I have gained from my extensive exposure to these equipment. During the four decades of my service in Tamil Nadu Electricity Board (TNEB) as an engineer, I served many fields such as distribution, EHV transmission, substation and transformer repair workshop. This gave me invaluable experience specifically in the repair and maintenance of transformers, and generally, in other equipment. Further, a masters degree in engineering with specialisation in power systems afforded me the opportunity of a long period of service as a faculty in TNEB Training Institute, and gave me the chance to interact with engineers and technicians on power transformer applications, a privilege I continue to enjoy as a visiting faculty of the institute.

The aim of this book is to provide practical information on power transformers that will be useful to practising engineers. It begins with a review of the theory of transformers required for making use of the practical information provided later on in the book. The book can be used by graduate and postgraduate students of Power System Engineering to access the application part of transformers.

I would consider my effort of putting together this useful and practical 'information provider' fruitful if it is used by the intended reader to augment the knowledge found in other books and manuals on power and distribution transformers. I welcome suggestions for improving the book further in this regard.

Acknowledgments

At the outset I wish to acknowledge my parent organisation, Tamil Nadu Electricity Board (TNEB)—where I worked for nearly four decades as Engineer in various field of Power System—for providing me the opportunity to acquaint myself with the various practical aspects of the system. This has provided me the motivation for writing this book. I owe my thanks to the former deputy directors of the Transmission and Substation Training Institute of TNEB, Shri B. Jeyaraman, ME, FIE, and Shri R. Pannerselvam, BE, MIE, for their support and encouragement of my venture.

I am indebted to the then Deputy Director of Training, Transmission and Substation Institute, Shri V. Nalendran, BE, MSc (Engineering), for the support, encouragement and the opportunities for exposure to power systems technical activities that has provided me the required thrust for preparing and completing this book. I also express my thanks to the present faculty members of the institute for their patronage and the regard accorded by them for my work.

Finally, the dream of publishing this book, conceived in my mind many years ago, would not have come to fruition but for my son Hariharan, who put in time and effort to bring it to a shape by keying-in the contents. Without his continued support this book would never have seen the light of day even though the materials were in some state of readiness for publishing. I would also like to thank Universities Press for taking up this work for publishing and Ms Sreelatha Menon in particular for editing and giving shape to the book.

H. Kailasaraman

1

Power and Distribution Transformers: Theory, Conventional Features and Design

This section reviews the basic theory of power and distribution transformers. The losses in transformers, transformer efficiency and equivalent circuits with per unit parameters and their use in the calculation of voltage regulation and power loss are discussed. Detailed descriptions of magnetic cores, windings and coil interconnections in windings are provided. Tap changing methods with on-load and off-load tap changers are explained.

1.1 Basic Functions of a Power Transformer

1.1.1 Introduction

Transformers employed in electrical power systems are classified into two categories—power transformers and instrument transformers. The main functions of power handling and voltage transformation are performed by power transformers while subsidiary functions such as metering and protection are performed by instrument transformers.

The nomenclature of power transformers varies according to the place of its usage. In the receiving station it is known as an *auto transformer*; in the substation it is called a *power transformer*; and in the distribution location it is called a *distribution transformer*.

The discussions in this section pertain to power transformers only. However, appropriate nomenclature is mentioned in places where it is specifically required.

1.1.2 Purpose of a Power Transformer

The power transformer serves the functional requirement of a power system (substation), namely transmitting the required power and transforming the same from one voltage level to another level according to requirements. The other equipment associated with the power system perform supportive functions such as feeding, switching and protection.

To minimise the amount of power loss in transmission, it is necessary for the power system to transmit power at EHV level and subsequently reduce the same to supply loads at the required voltage levels. Such a transmission system involves many substations with installations of power transformers. The power transformer transforms the voltages to the required level even while maintaining the amount of power transmitted the same with high voltage, low current (HV, LC) on one side and low voltage, high current (LV, HC) on the other side. Industries that avail services at EHT or HT level will have to employ their own transformers and associated equipments to draw the required operating voltage.

1.1.3 Main Parts of a Transformer

The parts of a transformer are shown in Fig. 1.1. There are two windings on a power transformer, one for high voltage and the other for low voltage. The coils have a certain number of turns, depending upon the voltages being handled. The windings are centred on a magnetic iron core, which is mainly responsible for the transformer action. This core-coil unit is placed inside a metallic tank and immersed in transformer oil which acts as an insulant and coolant.

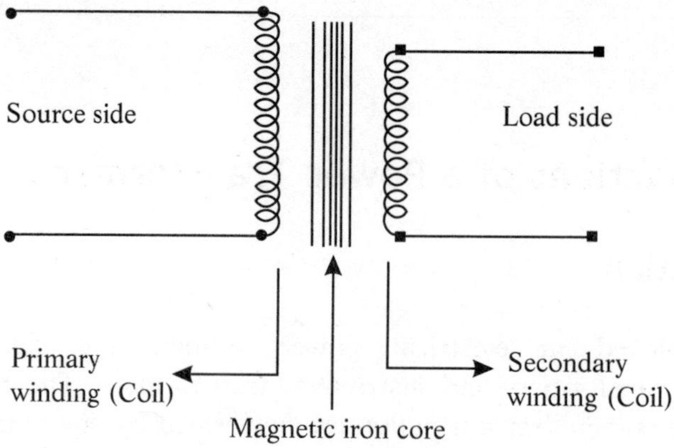

Fig. 1.1 *Parts of a single-phase transformer*

One of the transformer windings is connected to the source side while the other is connected to the load side. The windings on the source side and the load side are called the *primary winding* and the *secondary winding* respectively.

The transformer can function either as a *step-up* or *step-down* transformer, depending on the required voltage levels on the source side and the load side. In a step-up function, the primary will be at low voltage–high current while the secondary will be at high voltage–low current (as in a generating station); in a step-down function it is the converse of the above.

There are no physical electrical connections between the primary and secondary winding. They are linked only magnetically. Hence, physical connections from the supply and load sides end abruptly on their windings in a power transformer.

The auto transformer, however, is a special type of power transformer that has only a single winding and tappings on it for the HV side and LV side and where physical connections exist between the source and load sides. This special type of transformer is discussed separately in the book.

1.1.4 Transformer Action under No Load

Under no load, the following sequence of events take place.

(i) AC supply voltage is given to the primary winding.
(ii) Primary coil is energised. A small magnetising current flows in the winding. This quantity is dependent upon the impedance of the winding. The impedance being large, a very low current of the order of 1% of the full-load current will flow.
(iii) This AC magnetising current produces a magnetic field in the primary winding, which is also alternating. A back emf E_P is produced in the primary, which opposes the applied voltage V_P (an action symbolic of inductive reactance), and so the magnetising current is low.

Note: The moment V_P is applied, there will not be any E_P produced to oppose it, since the production of E_P will take some time after the application of voltage (of the order of milliseconds only). In the meantime, the resistance of the primary winding is the only limiting factor to reduce the current flow, and so there will be a very heavy current for a very small duration of time on energising the primary. This is called the *inrush current*. The inrush current will be of the order of 10 to 20 times the full-load current and will settle down to its low value when the effect of back emf E_P becomes pronounced.

(iv) The iron core of high magnetic permittivity circulates a magnetic field around it.
(v) This field links with the secondary winding, which is also wound on the same core.
(vi) The alternating magnetic field linking the secondary winding conductors produces an AC voltage E_S in that winding.
(vii) This induced voltage is collectively available at the open-circuited secondary terminals; the quantity depends upon the number of turns available in the secondary winding.
(viii) The secondary terminal is now ready to feed loads.

(ix) When loads are fed, the secondary terminal voltage is V_S, which is less than E_S due to voltage drop in the secondary winding by the load current flowing through the secondary winding. With no load, $V_S = E_S$.

The above sequence is shown in Figures 1.2(a) and (b).

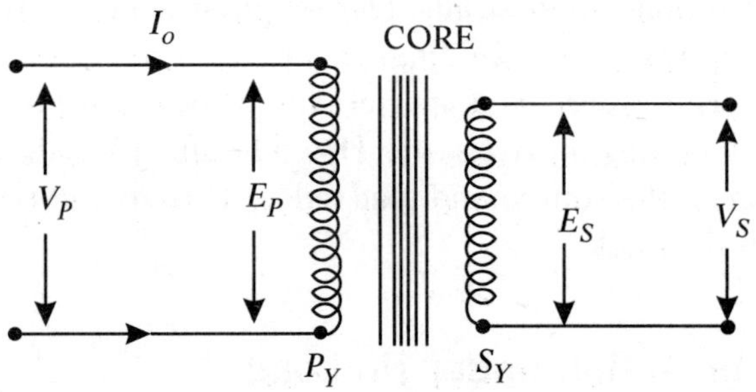

Fig. 1.2(a) *Schematic of transformer under no load*

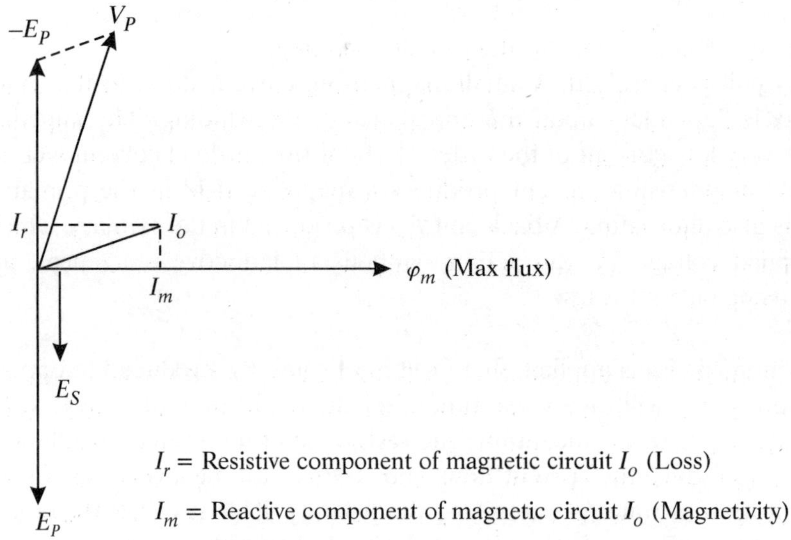

Fig. 1.2(b) *Vector diagram (no load)*

1.1.5 Transformer Action under Load

An AC voltage is made available in the secondary terminals by the transformer action. A load is connected to the secondary terminals. The sequence of events is listed below.

1. Load L with a power factor $\cos\varphi$ is connected to the secondary leads.
2. Due to the potential V_S, a load current I_S is supplied from the secondary. This current is, however, to be drawn only from the source side.

3. Hence, current is drawn from the source and primary current I_P flows. This I_P comprises the original magnetising current I_o and the current I_{PS} to compensate the secondary current I_S.
4. This primary current drawn from the source is automatically adjusted according to the fluctuating load requirement in the secondary.
5. In this process, the magnetising current of the transformer remains undisturbed, i.e., I_o is constant and only I_{PS} varies.
6. The power transferred to the load being the same, the values of I_{PS} and I_S depends on V_P and V_S respectively.
7. In the primary, $V_P > E_P$ and in the secondary, $V_S < E_S$ under loaded conditions.

The above sequence is shown in Figures 1.3(a) and (b).

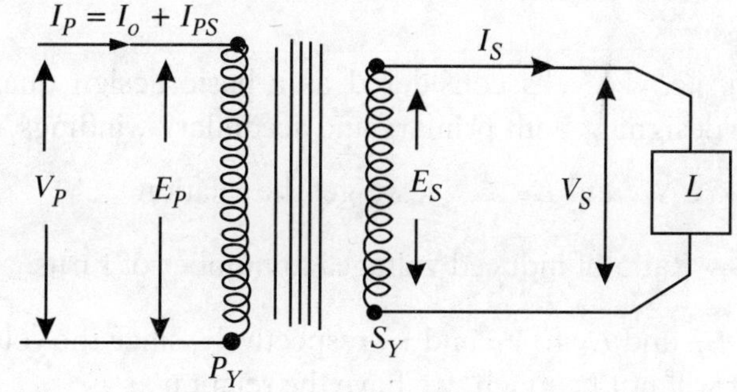

Fig. 1.3(a) *Schematic of transformer under load*

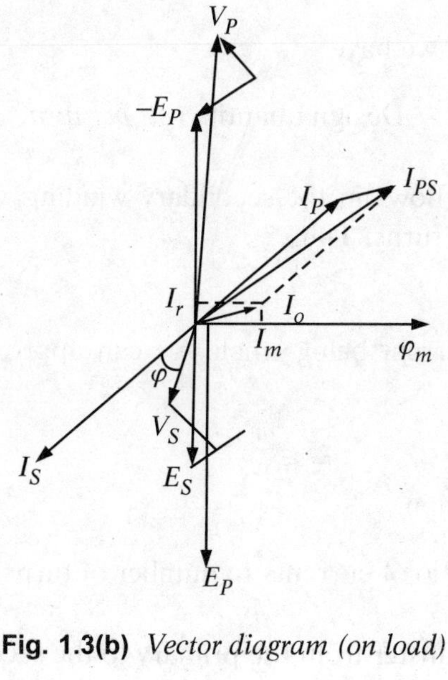

Fig. 1.3(b) *Vector diagram (on load)*

1.1.6 Relations of Ratios

(i) The voltages induced in a transformer winding is given by the formula

$$E = 4K_f \varphi_m N f,$$

where the form factor K_f, flux φ_m, and the frequency f are constants, both for the primary and secondary windings, whereas the number of turns N are different.

Let N_P be the number of turns in the primary and N_S the number of turns in the secondary.

The primary and secondary windings are wound on the same core and the magnetising flux linking the windings is the same. Obviously, the voltage induced in a single turn of each winding is the same. Therefore, the induced voltages in the primary and the secondary windings are in direct proportion to the number of turns in their windings.

Thus, volt/turn, i.e., V/N, is considered as a basic design quantity for a power transformer for designing both primary and secondary windings.

Hence, with $E_P \propto N_P$ and $E_S \propto N_S$, we get the relation

$$\frac{E_P}{E_S} = \frac{N_P}{N_S} \longrightarrow \text{Ratio of induced voltages to number of turns.}$$

Approximating E_P and E_S to V_P and V_S respectively, since the voltage drop through the transformer will not be much, we have the relation as

$$\frac{V_P}{V_S} = \frac{N_P}{N_S}.$$

Rearranging the above, we have

$$\frac{V_P}{N_P} = \frac{V_S}{N_S} = \frac{V}{N} \longrightarrow \text{Design quantity } \textit{volt per turn}.$$

(ii) When load current I_S flows in the secondary winding, I_{PS} gets itself adjusted to maintain equal ampere turns. Thus,

$$I_{PS} N_P = I_S N_S.$$

I_o, the magnetising current being small, we can approximate I_{PS} to I_P (since $I_{PS} + I_o = I_P$).

$$\therefore \quad I_P N_P = I_S N_S.$$

Hence, we get the relation

$$\frac{I_P}{I_S} = \frac{N_S}{N_P} \longrightarrow \text{Ratio of currents to number of turns.}$$

(iii) The apparent power transfer from the primary to the secondary remains the same.

Thus,

$$V_P I_P = V_S I_S.$$

Hence, we have the relation

$$\frac{V_P}{V_S} = \frac{I_S}{I_P} \longrightarrow \text{Ratio of voltage to current.}$$

This relation can also be obtained from the previous two relations.

1.1.7 Losses in Power Transformers

The losses in the transformer can be classified as (i) no-load losses and (ii) losses while supplying load. The no-load losses are called *constant losses,* while the losses in load are the *variable losses.*

(i) No-load losses

These losses are due to

(a) Hysteresis loss
(b) Eddy current loss
(c) Stray losses and
(d) Copper loss in primary due to magnetising current.

(a) *Hysteresis loss*: The magnetic core is subjected to reversal of magnetisation as many number of times as the frequency of supply. Negative maximum to positive maximum is the increasing trend; positive maximum to negative maximum is the decreasing trend. During both trends, the magnetisation and demagnetisation do not follow the same path but follows a closed cycle, leaving some residual magnetisation which is to be replenished (Fig. 1.4). This is called the *hysteresis loss*.

The magnetisation current being constant, this is a constant loss.

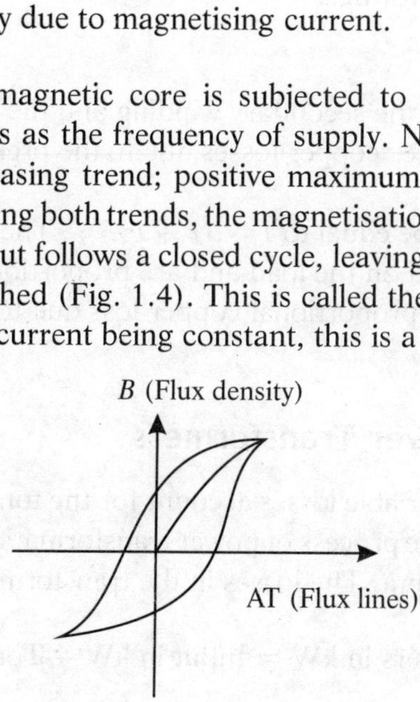

Fig. 1.4 *Magnetisation curve*

(b) *Eddy current loss*: The magnetic field cuts the core cross-section also. Being an iron core, a current called eddy current is circulated, spreading through the cross-section of the core. This current in association with the resistivity of the iron core produces heat and accompanying loss. The eddy current is proportional to the cross-section of the core.

The completion of the magnetic circuit in the core requires only the linear path and not the cross-section. Hence, the required cross-section of the core is assembled with laminations of very small thicknesses. A continuous huge cross-section is thus avoided since the faces of the laminations are insulated. The quantity of eddy current produced is thus very small, proportional to the smaller thickness of the laminations only.

The construction of the core without laminations will lead to very heavy eddy currents. The laminations help to limit the eddy current to a tolerable value so that core may be put into service with losses due to eddy currents minimised. The flux density in the core being constant, this eddy current loss is also a constant loss.

(c) *Stray losses*: There are many iron materials in the core-coil unit construction such as the core bolt, frames, metal tank, etc., with which stray magnetic fields interact to produce stray eddy current losses.

(d) *Primary copper loss due to magnetising current*: A small magnetising current I_r flows through the primary winding which has some resistance R_p. Hence, $I_r^2 R_p$ is the primary copper loss due to magnetising current, which is also accqunted as a constant loss in the transformer.

(ii) Losses due to load

The load current I_S flowing in the secondary winding and the corresponding I_{PS} flowing in the primary winding produces copper losses due to the presence of resistances in both the windings.

The losses due to load will be equal to $I_{PS}^2 R_P + I_S^2 R_S$. These losses are called variable losses since I_{PS} and I_S depend on the load and are proportional to it. Hence, if full-load copper loss is estimated, then proportional copper loss due to load can be found for any load.

1.1.8 Efficiency of Power Transformers

The constant losses and the variable losses account for the total losses in the transformer. This is the power lost during the process of power transformation through the transformer, reducing the active power output. The losses in the transformer are manifested as heat.

$$\text{Output of power transformers in kW} = \text{Input in kW} - \text{Total losses in kW}$$

$$\text{Efficiency} = \left(\frac{\text{Output}}{\text{Input}}\right) \times 100\%$$

The losses are minimised with proper design features. Moreover, the transformer being a static device, the power transformer can be made to operate at as high an efficiency as 99.9%. Maximum efficiency is achieved when *iron loss = copper loss.*

The load at which this maximum efficiency is achieved depends on many design factors and hence is not the same for all transformers and need not occur at maximum load only. Moreover, operation of transformers at maximum efficiency is not possible always.

1.1.9 Calculation of Efficiency at Any Load

$$\%\,\eta = \left(1 - \frac{\text{Variable loss} + \text{constant loss}}{\text{Input power in watts}}\right) \times 100$$

$$= \left(1 - \frac{y \times (LF)^2 + x}{s \times PF \times 10^3}\right) \times 100,$$

where

the rated capacity of the transformer	=	s kVA
power factor at which the load works	=	PF
working load factor to rated capacity	=	LF
no-load loss	=	x watts
full-load copper loss	=	y watts.

$$\%\text{ Efficiency } \eta = \left(\frac{\text{Output in kW}}{\text{Input in kW}}\right) \times 100$$

$$= \left(\frac{\text{Input} - \text{Losses}}{\text{Input}}\right) \times 100 = \left(1 - \frac{\text{Losses}}{\text{Input}}\right) \times 100.$$

Example 1

For a transformer with working conditions as

$s = 250$ kVA; PF $= 0.8$; LF $= 3/4^{\text{th}}$ load; $x = 120$ watts; $y = 500$ watts

$$\%\eta = \left(1 - \frac{\left(500 \times (3/4)^2 + 120\right)}{250 \times 0.8 \times 10^3}\right) = 99.73\%$$

1.1.10 Equivalent Circuit Representation

The resistances of the primary and secondary windings, R_P and R_S, the reactances, X_P and X_S, are parameters of the transformer. These are series quantities. Further, to account for no-load losses and magnetising ampere turns which are constant quantities, a resistance R_m and reactance X_m are considered in parallel across the primary voltage. R_m and X_m are however not taken into account for system calculations since their effect is very small. Also, there will be capacitances in the transformer, the effect of which is negligible and hence not considered.

The above parameters are represented in the equivalent circuit shown in Fig. 1.5.

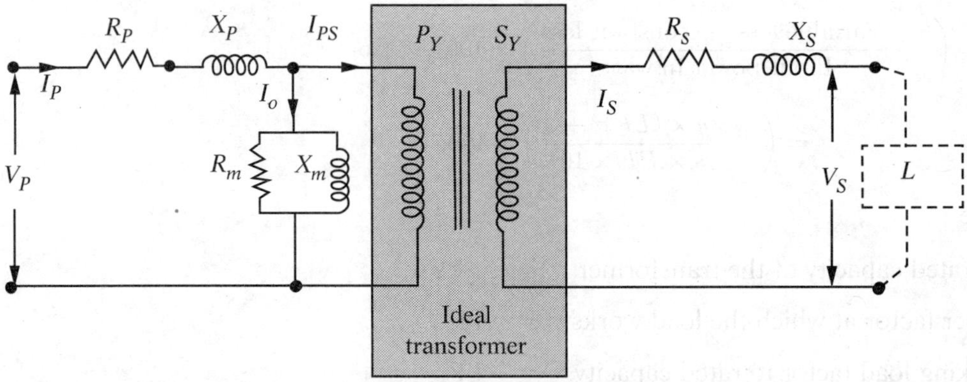

Fig. 1.5

1.1.11 Accounting for Parameters on One Side

Neglecting the parallel quantities, the series parameters can be transferred and accounted on one side with the other side being considered ideal. This is obvious because of the different voltage on both sides. This will help in representing the total impedance of the transformer as a whole for the entire transformer windings with reference to a particular voltage side, making system calculations easier.

(a) *Accounting for the parameters on the primary side with the secondary side kept ideal*

The equivalent circuit is shown in Fig. 1.6. The parameters as seen through the primary side of the transformer are represented as R_{PY} and X_{PY}, where

$$R_{PY} = R_P + R_S \left[\frac{N_P}{N_S}\right]^2 \text{ ohms}$$

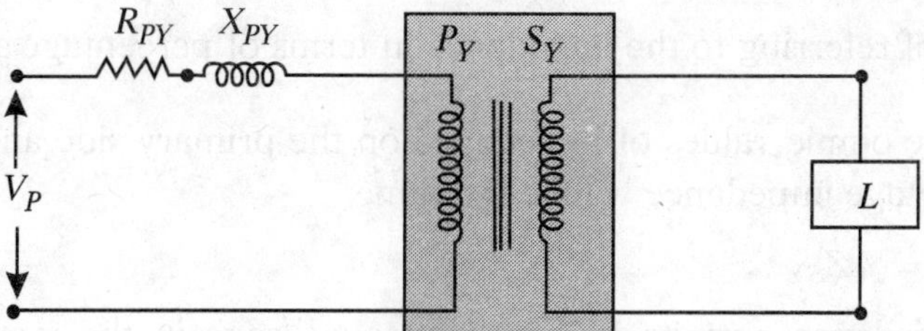

Fig. 1.6

and $X_{PY} = X_P + X_S \left[\frac{N_P}{N_S}\right]^2$ ohms.

Then, $Z_{PY} = \sqrt{R_{PY}^2 + X_{PY}^2}$

will be the total impedance in ohms seen through the primary side.

(b) *Accounting for the parameters on the secondary side with primary side kept ideal*

The equivalent circuit is shown in Fig. 1.7.

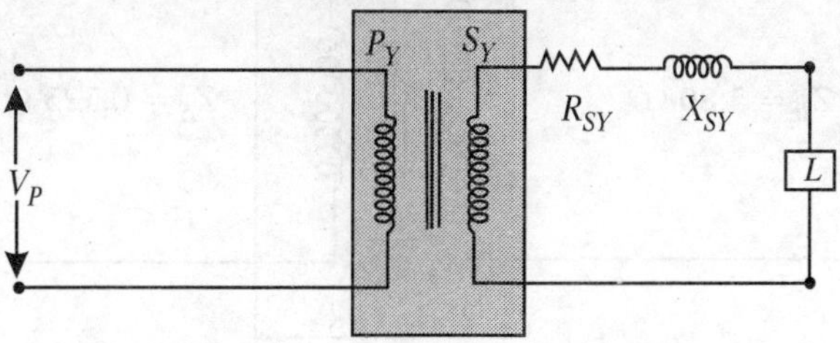

Fig. 1.7

Here, the parameters as seen through the secondary side of the transformer are represented as R_{SY} and X_{SY}, where

$$R_{SY} = R_S + R_P \left[\frac{N_S}{N_P}\right]^2, \text{ and } X_{SY} = X_S + X_P \left[\frac{N_S}{N_P}\right]^2.$$

Then, $Z_{SY} = \sqrt{R_{SY}^2 + X_{SY}^2}$.

1.1.12 Parameters Referred to as Percentage Values

The impedance of the transformer as seen through the primary side or the secondary side, when referred to in terms of percentages, is called the percentage impedance of the transformer. In all transformer usages, *percentage impedance* is one data that is denoted in the nameplate details of the transformer.

The advantages of referring to the impedance in terms of percentage are given below.

1. Whatever be the ohmic values of impedance on the primary side and the secondary side, the percentage impedance will be the same,

   i.e., $\%Z_{PY} = \%Z_{SY}$

2. Percentage impedance permits its homogeneous usage in the system calculations even though various voltage levels and various interconnections of the three-phase windings of the transformer are involved.
3. The equivalence of % impedances of various power transformers reveals the possibility for parallel operations, irrespective of their individual capacities. Such possibility is not obvious if only the ohmic values of impedances are known.

Percentage impedance calculation

Consider a single-phase transformer shown in Fig. 1.8. The respective impedances of the primary and secondary winding are shown in ohms.

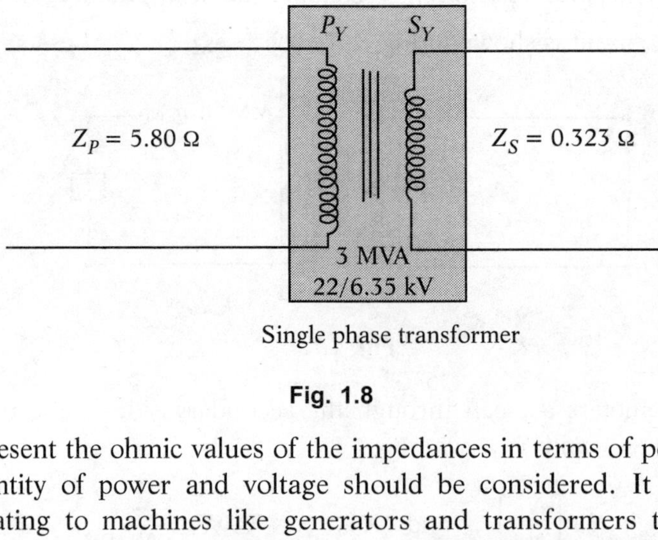

Single phase transformer

Fig. 1.8

In order to represent the ohmic values of the impedances in terms of percentage values, some base quantity of power and voltage should be considered. It is customary in calculations relating to machines like generators and transformers to consider their machine rating as the base. Here, 3 MVA is considered the *base power*, and the respective sides voltages, i.e., 22 kV or 6.35 kV, is considered the *base voltage*, depending on which side is being taken into consideration.

(i) If the primary side is considered, then

base kV	=	22 kV
base power	=	3 MVA
base impedance on the primary side	=	$\dfrac{(\text{Base kV})^2}{\text{Base MVA}}$
or $Z_{\text{base}(P_Y)}$	=	$(22)^2 \div (3) = 161.334\,\Omega$.

(ii) If the secondary side is considered, then

base kV = 6.35 kV

base power = 3 MVA

base impedance on the secondary side $= \dfrac{(\text{Base kV})^2}{\text{Base MVA}}$

or $Z_{\text{base}(S_Y)} = (6.35)^2 \div (3) = 13.440\ \Omega.$

Using the above base impedances, the % impedance of the transformer can be obtained by calculating the impedances of the primary (Z_P) and the secondary (Z_S) in any one of the sides as explained below.

(a) Accounting for the impedance on the primary side (see Fig 1.9a for details):

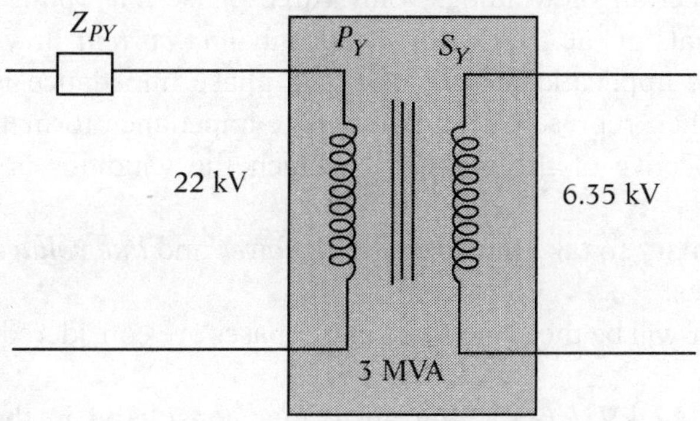

$$\begin{aligned}Z_{PY} &= Z_P + Z_S\,(N_P \div N_S)^2\\ &= 5.80 + 0.323(22 \div 6.35)^2\\ &= 5.80 + 3.88\\ &= 9.68\ \Omega\end{aligned}$$

$$\begin{aligned}\%Z_{PY} &= [Z_{PY} \div Z_{\text{base}(P_Y)}] \times 100\\ &= (9.68 \div 161.334) \times 100\\ &= 6.0\%\end{aligned}$$

$Z_P = 5.80\ \Omega$; $Z_S = 0.323\ \Omega$

Fig. 1.9(a)

(b) Accounting for the impedance on the secondary side (see Fig. 1.9b for details):

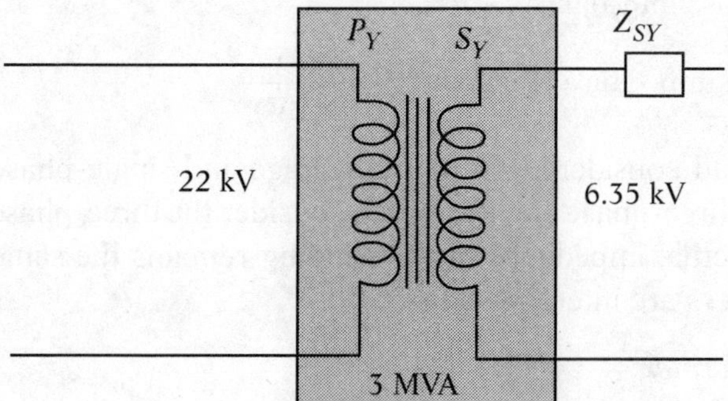

$$\begin{aligned}Z_{SY} &= Z_S + Z_P\,(N_S \div N_P)^2\\ &= 0.323 + 5.80(6.35 \div 22)^2\\ &= 0.323 + 0.483\\ &= 0.806\ \Omega\end{aligned}$$

$$\begin{aligned}\%Z_{SY} &= [Z_{SY} \div Z_{\text{base}(P_Y)}] \times 100\\ &= (0.806 \div 13.440) \times 100\\ &= 6.0\%\end{aligned}$$

$Z_P = 5.80\ \Omega$; $Z_S = 0.323\ \Omega$

Fig. 1.9(b)

Note: We thus find from the above example that $\%Z_{PY} = \%Z_{SY} = 6\%$ even though the ohmic values of Z_{PY} and Z_{SY}, 9.68 Ω and 0.806 Ω respectively, are different.

Thus, it confirms the advantage No. 1 referred to earlier.

1.1.13 Parameters in Percentages for Three-phase Transformers

So far the discussions were confined to single-phase transformers. The above treatment will hold good for three-phase transformers as well. A three-phase transformer has three different sets of single-phase windings around three independent core limbs. These core limbs are interconnected to form a common magnetic circuit. The phase windings are interconnected in delta, star or zigzag fashion, depending on the specific requirement. These will be discussed in detail in the forthcoming sections.

Whatever be the type of interconnection of windings, only three-phase *line voltage* is applied between the bushing terminals of the transformer and only *line current* flows through the terminals. In three-phase applications also, only per phase impedance is considered in all calculations. This, when represented in percentage impedance format, holds good for all calculations irrespective of the manner in which the windings are interconnected.

In a three-phase systems it is customary to take the *three-phase power* and *line voltage* of the respective sides as base quantities.

The % impedance of the transformer will be the same if the above bases are considered. This is explained below.

Consider the single-phase 3 MVA 22 kV / 6.35 kV transformer considered in the earlier example. Three such single-phase units are interconnected in delta in the primary and in star in the secondary so that the bank will be on 3-phase application for 22/ 11 kV voltage and 9 MVA power. We know the relation

$$\text{Percentage impedance} = \frac{\text{Actual impedance in ohms}}{\text{Base impedance}}$$

$$= \text{Actual impedance in ohms} \times \frac{\text{Base MVA}}{(\text{Base kV})^2}.$$

For the single-phase application we had considered the phase voltage and single-phase power as the base quantities. Now, for three-phase application we consider the three-phase power and line voltage as bases; the actual impedance of the winding remains the same since the same single-phase transformers are interconnected.

(i) On the secondary side of the transformer

The secondary side is in star and hence the actual impedance of the winding is in series with the line directly (see Fig. 1.10a).

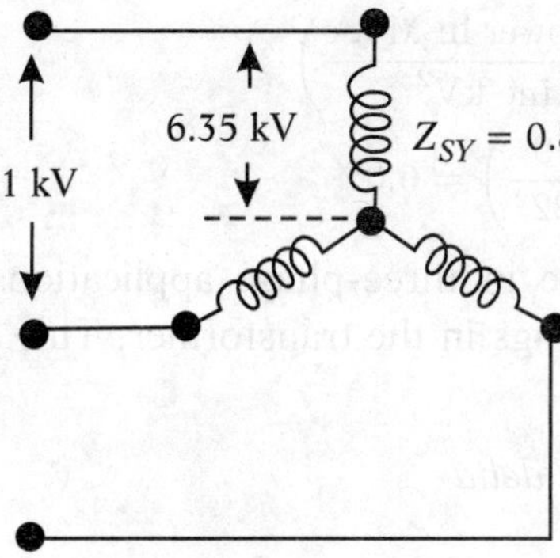

% Z_{SY} for 3 phase interconnection

$$= Z_{SY}\,\Omega \times \frac{\text{3 phase power in MVA}}{\text{line kV}^2}$$

$$= 0.806 \times \frac{9}{11^2}$$

$$= 6.0\%$$

Fig. 1.10(a)

(*ii*) *On the primary side of the transformer*

The primary winding is not in series with the line since it is interconnected in delta (Fig. 1.10(b)). Hence, the impedance of each phase winding cannot be taken directly as the per phase impedance, and an equivalent star should be found. The *per phase impedance* directly in series with the line should then be calculated and applied (see note below).

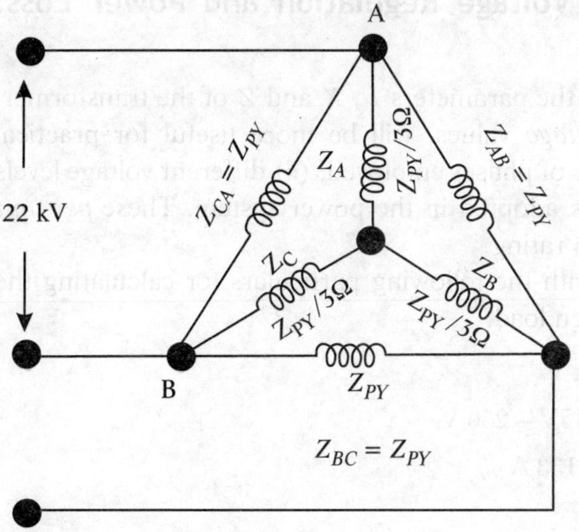

Delta connected primary with equivalent star shown inside

$Z_{PY} = 9.68\ \Omega$

Fig. 1.10(b)

The equivalent per phase star impedance is

$$\frac{Z_{PY}}{3} = \frac{9.68}{3} = Z_A \text{ (The formula for the calculation is given in the note below.)}$$

$$\% \ Z_{PY} \text{ for } 3\varphi \text{ interconnection} = Z_A \times \left(\frac{3\varphi \text{ power in MVA}}{\text{Line kV}^2}\right)$$

$$= \left(\frac{9.68}{3}\right) \times \left(\frac{9}{22^2}\right) = 6.0\%$$

Hence, we find percentage impedance is the same in three-phase applications also, whatever may be the interconnection of phase windings in the transformer. This proves the advantage No. 2 referred to earlier.

Note: *Formula for calculation of equivalent star from delta*

From Fig 1.10(b),

$$Z_A = [Z_{AB} \times Z_{CA}] \div [Z_{AB} + Z_{BC} + Z_{CA}].$$

Since $Z_{AB} = Z_{BC} = Z_{CA}$ for balanced circuit, we have

$$Z_A = [Z_{PY} \times Z_{PY}] \div [Z_{PY} + Z_{PY} + Z_{PY}]$$

$$= [Z_{PY} \times Z_{PY}] \div [3Z_{PY}] = Z_{PY}/3\Omega.$$

1.1.14 Use of Parameters for Voltage Regulation and Power Loss: Sample Calculation

It is seen from the previous sections that the parameters *R*, *X* and *Z* of the transformer, when represented in *per unit* or *percentage* values, will be more useful for practical applications irrespective of (i) the number of phases employed, (ii) different voltage levels and (iii) mode of phase interconnections adopted in the power system. These *per unit* values are based on the transformers own rating.

Consider a three-phase transformer with the following particulars for calculating the *power loss* and *voltage regulation* at a given load.

Capacity	:	100 kVA
Voltage ratio	:	11kV/433V – 250 V
Current ratio	:	5.25 A / 133 A
Percentage impedance	:	5%
Vector group	:	4_1 DY_{11}
No-load loss	:	180 watts
Full-load loss	:	1600 watts

The parameters which are considered as *base* for per unit calculation and *per unit values* of other parameters calculated with these bases are listed in Table 1.1.

The power loss and efficiency [η] already described in the previous sections may be recalled for representation through per unit for the transformer now under consideration.

Table 1.1 *Base parameters for per unit calculations*

Sl No.	Parameters	Quantity	Per unit value	% value	How arrived at
1	Three-phase capacity of power	100 kVA 100 kW	0.1	10%	Base adopted (capacity in MVA = 0.1 p.u.)
2	Line voltage HV LV	 11 kV 433 V	 1.0 1.0	 100% 100%	Base quantity of line voltage
3	Phase voltage HV LV	 11 kV/$\sqrt{3}$ 250 V	 1.0 1.0	 100% 100%	Base quantity of phase voltage
4	Line current HV LV	 5.25 A 133 A	 1.0 1.0	 100% 100%	Base quantities of line currents
5	Base impedance HV	1210Ω	1.0	100%	$\frac{\text{Base kV}^2}{\text{Base MVA}} = \frac{(11)^2}{0.1}$ or $\frac{\text{Phase voltage}}{\text{Line current}} = \frac{11\text{ kV}/\sqrt{3}}{5.25}$
6	Base impedance LV	1.88Ω	1.0	100%	$\frac{\text{Base kV}^2}{\text{Base MVA}} = \frac{(0.433)^2}{0.1}$ or $\frac{\text{Phase voltage}}{\text{line current}} = \frac{250}{133}$
7	Impedance as seen from HV side (Z_{HV})	60.50 Ω	0.05	5%	$\frac{\text{Actual impedance}}{\text{Base impedance}} = \frac{60.5}{1210}$
8	Impedance as seen from LV side (Z_{LV})	0.094 Ω	0.05	5%	$\frac{\text{Actual impedance}}{\text{Base impedance}} = \frac{0.094}{1.88}$
9	Resistance as seen from LV side(R_{LV})	0.030 Ω	0.016	1.6%	Calculated from full-load loss or by measurement $\Big\} = \frac{0.030}{1.88}$
10	Reactance as seen from LV side (X_{LV})	0.090Ω	0.0474	4.74%	$\sqrt{Z_{LV}^2 - R_{LV}^2}$ or $= \frac{0.090}{1.88}$ by measurement

The kW power supplied by the transformer as demanded by the load is to be reckoned with the load current and power factor of the load. Hence, the percentage of load and the power factor (PF) of the system will have to be considered for the power loss and voltage regulation calculations at appropriate places.

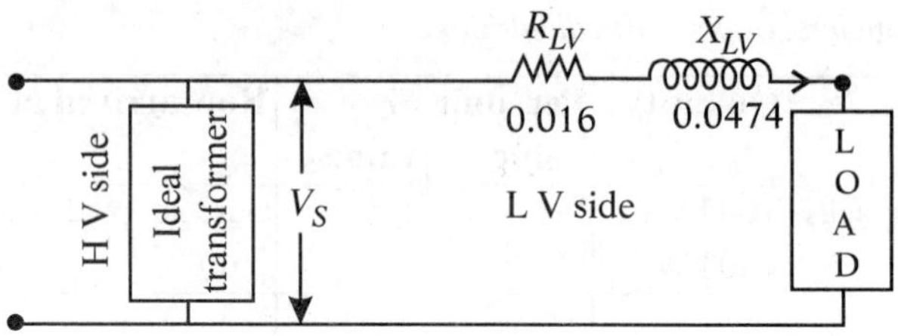

Fig. 1.11

The equivalent circuit for the above calculations is shown in Fig. 1.11 with per unit parameters.

Power loss at any load

(i) Input power in kW $= \text{Rated power} \times \left(\frac{\%\text{Load}}{100}\right) \times \text{P.F}$

$= 1 \times \left(\frac{\%\text{Load}}{100}\right) \times \text{P.F}$

(ii) Load losses $= \frac{\%\text{Load}^2}{100} \times (I_{FL}^2 \times R_{LV} \times 3)$, where I_{FL} is full-load current

(iii) No-load losses = Given (or) obtained by conducting core loss test

(iv) Total losses = No-load loss + Full-load loss

(v) Efficiency η in % $= \left(\frac{\text{Input power} - \text{Total losses}}{\text{Input power}}\right) \times 100$

Note: The load losses at full load can also be obtained by inserting a wattmeter during short circuit tests as described in Section 2.2.4. The reading of a wattmeter in a phase multiplied by 3 will give the total losses in watts.

Now, in the case of transformer considered in this example:

Rated kVA = 100 kVA = 1 p.u.

No-load loss = 180 watts $= \frac{0.180}{100}$ = 0.0018 p.u.

$I_{\text{Full-load}}$ LV = 133 A = 1 p.u.

R_{LV} = 0.030 Ω = 0.016 p.u.

(a) Calculation of power loss and efficiency (η) for 75% of load at 0.8 PF

Input power $= 1 \times 0.75 \times 0.8 = 0.6$p.u.

Load losses $= [0.75]^2 \times [1^2 \times 0.016 \times 3] = 0.027$p.u.

No-load loss = 0.0018 p.u.

$$\text{Efficiency } \eta = \frac{0.6 - (0.0018 + 0.0270)}{0.6} \times 100$$

$$= \frac{0.6 - 0.0288}{0.6} \times 100 = 95.2\%$$

(b) Calculation of power loss and efficiency (η) at full load at unity PF

Input power $= 1 \times 1 \times 1 = 1$ p.u.

Load losses $= 1^2 \times [1^2 \times 0.016 \times 3] = 0.048$ p.u.

No-load loss $= 0.0018$ p.u.

$$\text{Efficiency} = \frac{1.0 - (0.0018 + 0.048)}{1.0} \times 100$$

$$= \frac{1.0 - 0.0498}{1.0} \times 100 = 99.502\%$$

It may be noted that % efficiency varies at various loads with various PFs.

Voltage regulation at any load

The no-load voltage at the secondary output of the transformer, say V_{SO}, will be maintained at the rated voltage, say 1.0 p.u. (i.e., 433 / 250 V). The voltage at the same point at loaded condition will be less, say $V_{S\text{load}}$. The voltage regulation of the transformer is then defined as

$$\%\ \text{VR (Voltage Regulation)} = \frac{V_{SO} - V_{S\ \text{load}}}{V_{S\ \text{load}}} \times 100.$$

In the above, ($V_{SO} - V_{S\ \text{load}}$) is the voltage drop inside the transformer due to load. This drop in voltage is due to the flow of load current with a power factor over the R_{LV} and X_{LV} parameters of the equivalent circuit. This load current is proportional to the load.

The I_Z drop cannot be taken as it is since vectoral multiplication and addition is involved due to the presence of power factor. This 'drop' can be represented in a vector diagram. The complex algebraic method is given below.

Any load current I with a PF angle 'φ' can be represented by $I(\cos\varphi - j\sin\varphi)$. The drop due to this flow through R and X of the circuit will be

$$I[\cos\varphi - j\sin\varphi] \times (R + jx)$$

$$= I(R\cos\varphi - jR\sin\varphi + jx\cos\varphi + x\sin\varphi)$$

$$= I[(R\cos\varphi + x\sin\varphi) + j(x\cos\varphi - R\sin\varphi)].$$

Now, considering the current in terms of full-load current I_{FL} and percentage of load, the equation can be formed as:

$$\text{Voltage drop} = \frac{\text{\% of Load}}{100} \times I_{FL} \times [(R\cos\varphi + x\sin\varphi) + j\,(x\cos\varphi - R\sin\varphi)].$$

Voltage at load = $V_{SL} = V_S -$ Drop.

This also is a vectoral subtraction as may be seen in vector diagram, and hence,

$$V_{SL} = \left[V_S - \frac{\text{\% of Load}}{100} \times I_{FL} \times (R\cos\varphi + x\sin\varphi)\right] - j(\cos\varphi - R\sin\varphi).$$

Since, V_S and I_{FL} are 1.0 per unit, it may be rewritten as

$$V_{SL} = [\,1 - \frac{\text{\% of Load}}{100} \times (R\cos\varphi + x\sin\varphi)\,] - j(x\cos\varphi - R\sin\varphi).$$

Calculation of voltage regulation for full load with unity PF ($\cos\varphi$ = 1 and $\sin\varphi$ = 0)

$$\begin{aligned} V_{SL} &= [1 - (0.016 + 0)] - j(0.0474 - 0) \\ &= (0.984) - J(0.0474) = 0.9705 \text{ p.u.} \\ &= 0.9705 \times 250 \text{ volts} = 242.6 \text{ volts.} \end{aligned}$$

$$\begin{aligned} \text{\% Regulation} &= \frac{1 - 0.9705}{0.9705} \times 100 = 3.04\% \\ \text{Drop} &= V_{SO} - V_{SL} \\ &= 250 - 242.6 = 7.4 \text{ volts.} \end{aligned}$$

That is, in this loaded condition there is a drop of 7.4 volts inside the transformer and the voltage output at transformer secondary will be 242.6 volts only.

Calculation of voltage regulation for 50% load with 0.8 PF ($\cos\varphi = 0.8$ and $\sin\varphi = 0.6$)

$$\begin{aligned} V_{SL} &= [1 - 0.5(0.016 \times 0.8 + 0.0474 \times 0.6)] - j(0.0474 \times 0.8 - 0.016 \times 0.6) \\ &= [1 - (0.0064 + 0.01422)] - j(0.03792 - 0.0096) \\ &= (0.097938 - j0.02832) \\ &= \sqrt{0.979382 + 0.028322} \\ &= 0.9798 \text{ per unit.} \end{aligned}$$

V_{SL} in volts = $0.9798 \times 250 = 244.95$ volts.

Voltage drop, in per unit = $V_{SO} - V_{SL} = 1.0 - 0.9798$.

Voltage drop, in volts = $250 - 244.95 = 5.05$ volts

$$\% \text{ Regulation} = \frac{(1 - 0.9798)}{0.9798} \times 100 = 2.062\%.$$

1.2 Magnetic Core

1.2.1 Core Formation

The magnetic core forms the basic structure of the transformer. The primary and secondary windings are wound around this core. The alternating magnetic flux concentrates on the core, which links with the windings inducing emfs in them. In this section, the core formation, mechanical arrangement for rigid fixing, clamping, seating and lifting are discussed. Constructional features and repair practices are also described.

The core should not be made of a solid steel section. If made so, the flux associated with this core will produce high eddy currents, making it impossible for the transformer to function. The production of eddy current is proportional to the area of cross section of the core. The eddy current flow in the core will result in power loss, which is manifested as heat. Hence, while forming the core it is necessary that eddy current and the consequent power loss be made as low as possible. In order to achieve this, instead of forming the core with a solid section, the required cross section is achieved with an assembly of sheet steel stampings or laminations of very small thickness, with varnish or other insulations coated on either face of the laminations.

The advantage of this type of formation is:

(i) The required number of laminations is stacked to obtain the required weight of iron and cross sectional area.

(ii) The thin film of insulation between laminations reduces axial width of the core prone to eddy current flow across it. Thus, only a very small thickness is available for the eddy current to flow, thereby reducing its magnitude to a minimum level.

The insulation used for the laminations must be completely oil proof and heat resistant. The thickness of each stamping should be in the range of 0.30 mm to 0.50 mm. This cannot be reduced further, since the plates become then mechanically weak, making them liable to buckling in handling.

The core consists of a *limb* around which the windings are placed. Thus, there will be 3 limbs for a three-phase transformer. These limbs are interconnected with a *yoke*, which serves to complete the magnetic circuit. The laminations used for assembling the limbs are called *limb laminations* and that used for yoke are called *yoke laminations*. The interconnection of the limbs are made both at the top and the bottom and are respectively called *top yoke* and *bottom yoke*.

The laminations are manufactured by cold rolling process using steel with a very small quantity of silicon alloy with a low carbon content. This material has low hysteresis loss and high permeability. An increase of silicon content results in high resistivity and low saturation flux density. During the cold rolling process, the crystals or the grains of the materials orient in the direction of rolling. Hence, this is called *cold-rolled grain-oriented silicon steel (CRGS)*.

If this steel is magnetised in the direction of orientation of its grains, it exhibits good magnetic properties, whereas if it is magnetised in the opposite direction, the magnetic properties are unfavourable.

Hence, special care must be taken in the design of corners in built-up cores so that the regular path of magnetic lines is maintained in the grain-oriented direction.

1.2.2 Types of Cores

Magnetic cores are classified as shell type and core type. The shell type magnetic circuit is branched. There are two yokes per limb that encircle the limb on both sides. Therefore, as the magnetic flux leaves a limb, it branches off in two paths. The cross-sectional area of the limb will be twice that of the encircling yokes. Hence, the mass of iron required for this type of core will be more and hence its use is limited to small capacity transformers only.

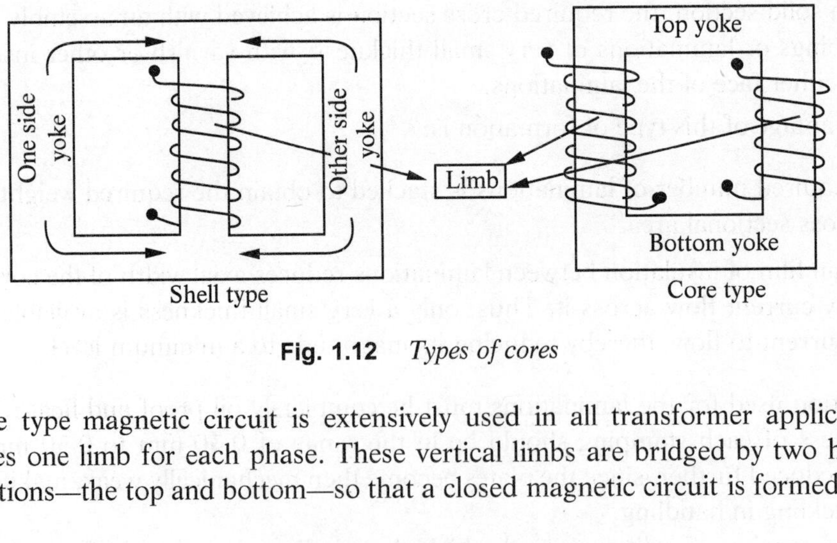

Fig. 1.12 *Types of cores*

The core type magnetic circuit is extensively used in all transformer applications. It comprises one limb for each phase. These vertical limbs are bridged by two horizontal yoke sections—the top and bottom—so that a closed magnetic circuit is formed.

1.2.3 Formation of Core-type Magnetic Circuit

The yoke laminations must be interleaved with limb laminations to enable release and insertion for repairs. Hence, the core can be formed through joints only. The yoke laminations will have joints with limb laminations. Thus, yoke and limb laminations cannot be jointly framed as a single lamination. The various types of jointed cores are discussed below.

(a) Butt joint core

Here, the top and bottom yokes are stacked separately. The whole yoke assembly is then butt-jointed with the limbs and clamped with vertical tie rods. So, in this type of jointing, the laminations need not be interleaved and joined individually. Hence it is easy to assemble, but this type of construction suffers from many drawbacks and is no longer in use today.

(b) Imbricated core

Unlike the butt-jointed core, the individual limb and yoke laminations are joined by interleaving or imbrication. There are two types of imbrications, and the manner by which the jointing is made specifies the types.

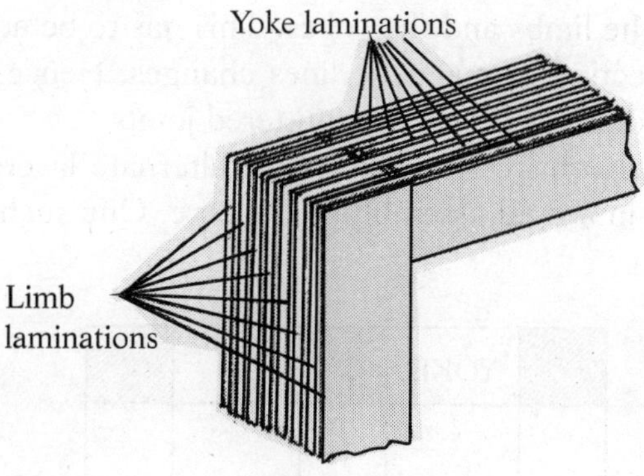

Fig. 1.13 *Imbricated core with right angled joints — isometric view*

(*i*) *Right-angled joints*: This type of core is built of interleaved construction of laminations in each layer of core making right-angled joints. The limb and yoke laminations are interleaved in a different manner in each layer. The short limb laminations are butt-joined with long yoke laminations in one layer, while in the next layer, long limb laminations are butt-joined with short yoke laminations, which will overlap the joint already made in the preceding layer. Such overlapping ensures rigid assembly of the core, (Fig. 1.14).

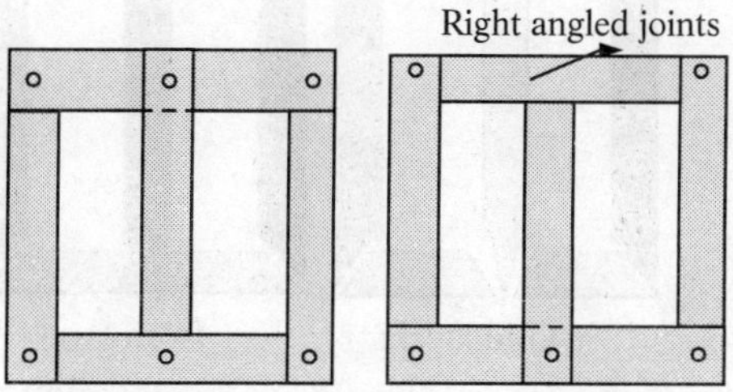

Fig. 1.14 *Arrangement of laminations*

By stacking layers over layers of such alternatively arranged laminations, a core of required thickness is obtained. This sort of imbricating is called *interleaving with right angled joints between laminations*. This type of right-angled interleaving is done for cores made of hot-rolled steel stampings only, where the direction of magnetisation is not a criterion.

(*ii*) *Bevelled or mittered joints*: In the magnetic core there are areas where the magnetic field passing from the limbs through the yoke and back changes its direction. This is shown as hatched areas in Fig. 1.15. In hot-rolled steel, this does not matter. But in cold-rolled steel, magnetisation in the rolled directions alone gives very good magnetic properties. Hence, in order to utilise the properties effectively, the cores are designed and assembled in such a way as to ensure that the lines of magnetic flux coincide with the steel rolling direction in the limbs and the yokes. This has to be achieved in the hatched areas also, where the direction of magnetic lines changes. Hence, the jointing of limbs and yokes laminations is done by bevelled or mittered joints.

Here also the different forms of interleaving in alternate layers provide overlapping over the joints and result in a rigid assembly of the core. One such interleaving is shown in Fig. 1.16.

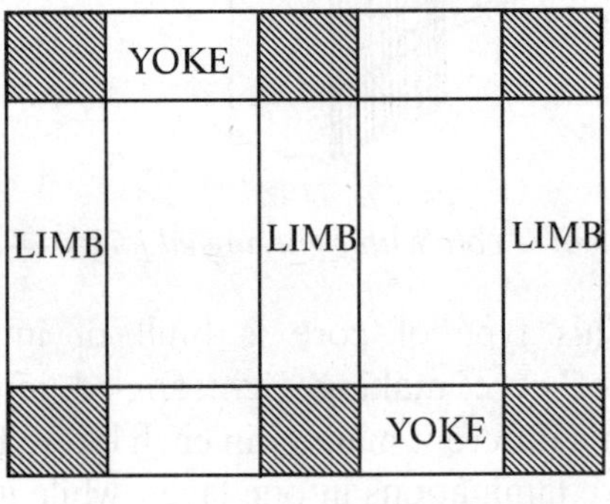

Fig. 1.15 *Arrangement of laminations*

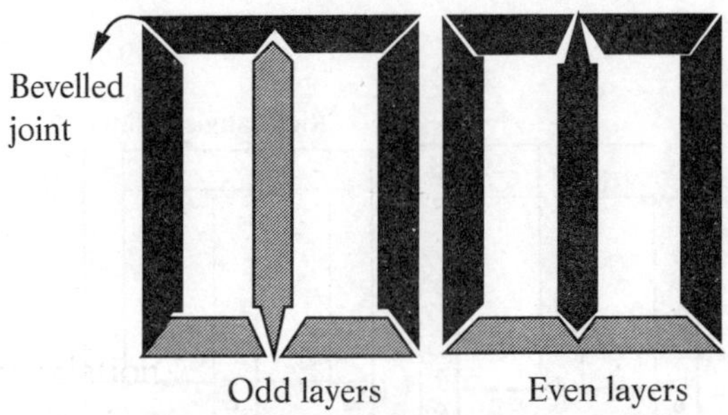

Fig. 1.16 *Forms of interleaving*

This type of mittered joint has the following advantages.

1. It reduces the magnetic circuit area where the lines of magnetic force do not coincide with the steel rolling direction.
2. It reduces the excitation current about 25 to 30%.
3. It reduces the no-load losses about 10 to 12%.

But fabrication of such laminations and assembling of core is complicated. This is, however, inevitable for CRGO steel used as core in transformers.

Be it a right-angled joint or mittered joint, if one single lamination plate is used for each layer, there will not be mechanical rigidity. Hence, two or three plates are used for each layer, and such interleaving is called double lamination interleaving or triple lamination interleaving. This method minimises the risk of imperfect interleaving due to buckling of plates compared to single plate interleaving.

1.2.4 Formation of Limb

The limbs are built to have a circular section. The advantages of employing a circular section rather than a rectangular section are as follows.

1. A rectangular winding wound on the core will have insufficient strength in the event of a short circuit.
2. A circular winding accommodated in a rectangular section will not fit in properly and will have a larger circumference. Therefore, the mean length of the turn will be more. With a circular core, the mean length of turn is reduced and there is savings in winding.

A perfect circular limb section requires each lamination with varied dimension and is hence uneconomical. The assembling of it too will be laborious. Hence, the core section is arranged in steps. Six or seven steps are made in the width of the limb to form a circular limb. The laminations of each step will have the same dimensions, but will vary between one step and another.

The steps are arranged in such a way that:

- The net sectional area is maximum for the number of steps employed.
- The corners of various steps lie on a circle of predetermined diameter.
- Increasing the number of steps will tend to realising as near a circle shape as possible.

Figure 1.17 shows the circular formation of core limb. The central step has the largest width, with the width of other steps slowly reducing. The length also reduces on either side along the centre line.

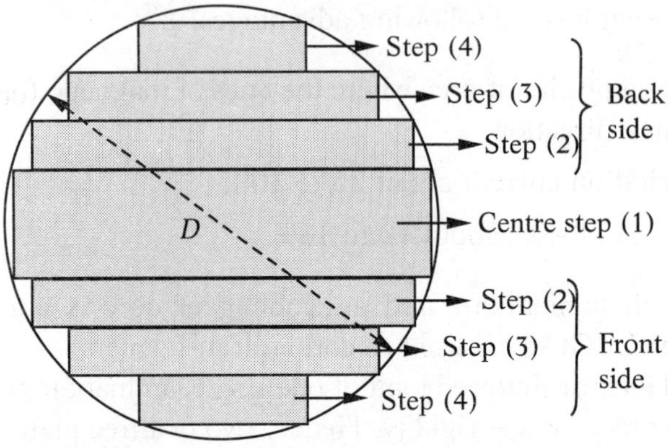

Fig. 1.17 *Plan of core limb*

1.2.5 Formation of the Yoke Section

The yoke section is arranged like the limb section so that the yoke laminations suits those of the limb section. The dimensions of the various sections of the yoke are so arranged that the sectional areas of each portion bear the same relation with the corresponding portion of the limb. This is important, as otherwise there will a tendency of cross fluxing between various sections causing excessive eddy currents.

1.2.6 Core Clamping

The core laminations should not be kept loose. The limbs and yokes are to be clamped tightly and proper compression is to be made so as to achieve

- the required electromagnetic characteristics of the core,
- proper mechanical strength,
- reduced hum due to magnetic vibration.

In small capacity transformers the core limbs need not be clamped because they are sufficiently rigid to ensure a stable vertical position of the core. Hence, while fitting the windings, the limbs are temporarily clamped with screw clamps. After fitting the windings, the required clamping tightness of limbs is ensured by wedging them with beachwood blocks or cleats.

In high capacity transformers, the clamping can be effected with or without clamping studs. Clamping with through studs is employed in cores made of hot-rolled steel.

Clamping with studs

Holes are punched in the limb and yoke laminations to fit the clamping through studs after the assembling of the core. The studs are to be insulated from the active steel and from the yoke clamps by means of paper base laminate tubes or fibre tubes and insulating washers made of press board, paper laminate or fibre.

The windings placed around the core limbs with wedging help in achieving uniform tightness in the limb. But, for the top and bottom yokes, channels are placed on either side of each yoke and clamping is done by through studs fitted at the points of intersection of the yoke and the limb axis, and thus a uniform tightness is ensured on the yoke. The yoke clamps may also be made of welded steel strips with required projections for fixing tie rods.

In addition, these yoke clamps support the windings and take up loads from lifting rods fitted in the respective holes on the yoke channels and serve to remove the core-coil unit from the tank. The top and bottom yokes are mechanically connected by means of vertical tie rods, which are also inserted in paper laminate tubes. These tie rods also serve the purpose of providing vertical clamping of the winding.

Beachwood blocks and steel strips secured to the bottom yoke clamp form a support for installing the core-coil unit inside the tank. The top yoke channel is also mechanically clamped to the tank through bolts and nuts. Thus, mechanical rigidity of the core-coil unit placed inside the tank is ensured during transport.

There should be insulation between the yoke channel and laminations. In large capacity transformers, there should be ducts wherein oil flow is established to dissipate the heat generated in the core due to hysteresis and eddy current losses. The cooling surface area is increased by forming longitudinal ducts in the active steel. Ducts are also formed between the yoke and yoke clamps by fitting vertical pressboard strips on to the insulation strip available at the inner side of the yoke clamp.

Details about core fixing and tightening arrangement with through studs and the complete core arrangements is shown in Figure 1.20. The threads of studs should be punched at three locations near the tightened nut to prevent loosening due to vibration during working.

Clamping without through studs

In case of cores made of cold-rolled steel, only this method should be used. The punched holes for through studs will be a hindrance to the magnetic lines of force, wherein the magnetic lines are made to bypass them and thus deviate from the rolling direction. Hence, clamping with through studs is not advisable.

The limb laminations are held together by stout webbing tape applied tightly on the smaller cores or by impregnated glass-fibre bands, which are effectively secured in position by the heat treatment of resin after application on the core. Glass fibre bands are also used in yoke clamping.

1.2.7 Earthing of Transformer Core

During the operation of transformer, the core and other metal components are subjected to a strong electric field by which they become charged. But individual components are charged differently. Therefore, there may be a potential difference between them which may cause sparking. To prevent this, the core and yoke clamps are earthed as shown in Fig. 1.18.

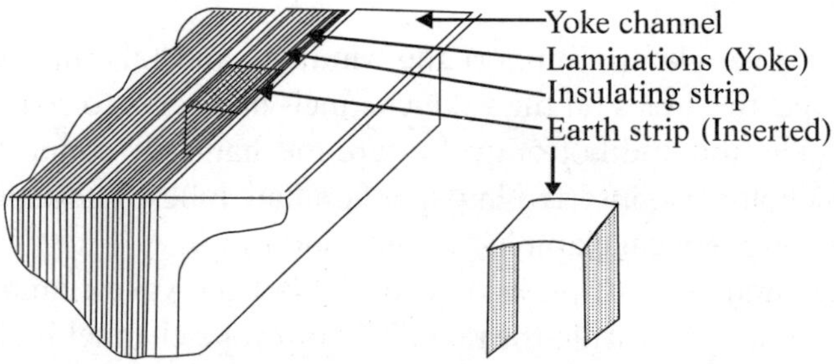

Fig. 1.18 *Earthing of the transformer core*

A tinned copper strip is inserted to a depth of 50 to 60 mm in the yoke section at a distance of 15 to 20 mm from the yoke clamp. The other end of the strip is clamped between the yoke channel and its insulating strip with the lamination, near the LV lead side. This earthing circuit is connected to the earthed main tank either through special earthing connections inside the tank or through lifting rods.

1.2.8 Basic Core Dimensions

The basic core dimensions of the transformer are shown in Fig. 1.19. This indicates the limit of dimensions of the coils to be rewound and the thicknesses of insulations and packing to be provided during assembly and repairs.

1.3 Windings

1.3.1 Introduction

The primary and secondary windings in a transformer perform the task of receiving the power available at one voltage level and transferring the same to the load at the required level with the help of the core. Each phase will have these sets of windings. The position of these windings in the limb, their choice and design considerations, winding directions, types of windings employed, insulations used etc., are discussed next in detail.

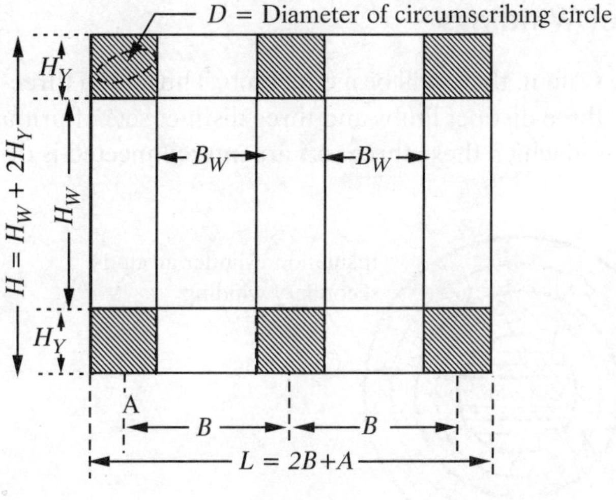

Fig. 1.19

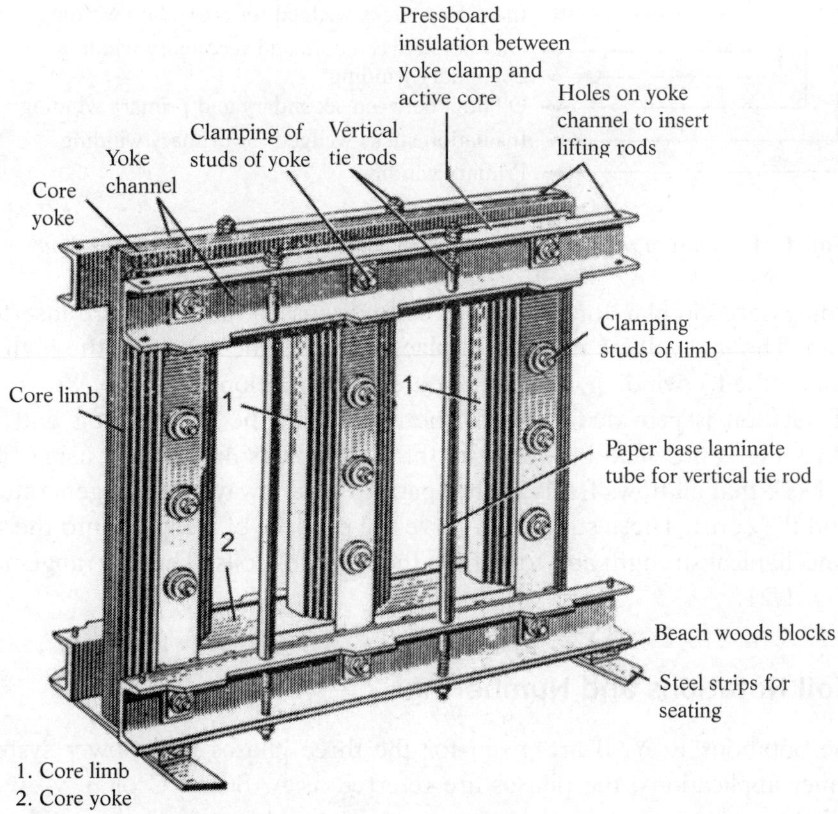

Fig. 1.20 *Assembled core with core bolt*

1.3.2 Position of Windings

For each phase of the system, there will be a core limb. Thus, for a three-phase system, the transformer will have three distinct limbs and three distinct sets of primary and secondary windings. The manner in which these three sets are interconnected is discussed in Section 1.4.

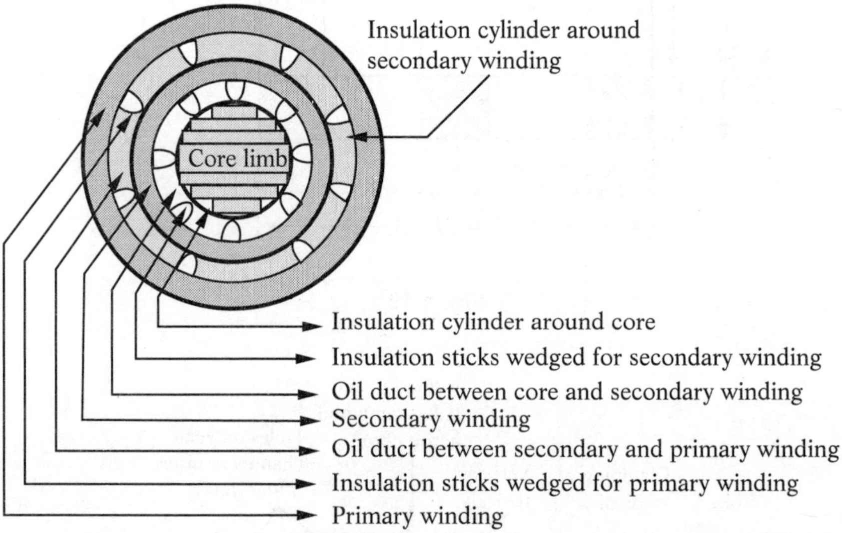

Fig. 1.21 *Plan of primary and secondary winding wound around core limb*

The windings are circular for reasons already discussed, and they are inserted over the core limb. The low voltage winding is placed around the core and the high voltage winding around the LV winding, i.e., in the outermost position.

Proper insulation is provided between the core limb, the LV winding and the HV windings. Spacing is provided between the three and ducts are formed using fibre and wooden sticks so that oil flows freely in the space to carry away the heat generated in the windings and the cores. These sticks also serve the purpose of wedging into the space to give more mechanical strength and to tighten the fixing of coils. These arrangements are shown in Fig. 1.21.

1.3.3 Coil Notations and Numbering

Usually, the notations R, Y, B are given for the three phases in a power system. But in transformer applications, the phases are referred as A, B and C or u, v, w. Hence, the phase limbs and their associated primary and secondary windings are also named similarly. Uppercase letters indicates primary winding while lowercase refer to secondary winding. The leads or connections brought out from the windings are denoted by suffixes of the concerned phase notation. Usually, lead numbers will start from the bottom and end at the top where it is connected to the bushings. A, B and C are referred from left to right when seen from the HV (primary) side.

Example 1: A representation of an A phase HV winding with six numbers of tapping leads is shown in Fig. 1.22.

Example 2: A representation of an A phase LV winding with top and bottom leads alone is shown in Fig. 1.23.

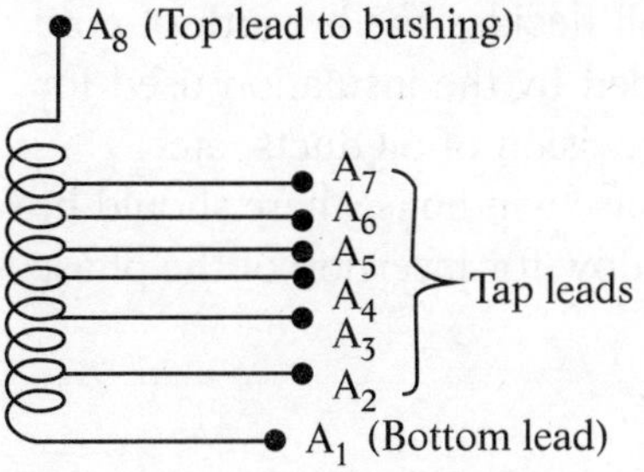

Fig. 1.22 *HV coil A phase limb*

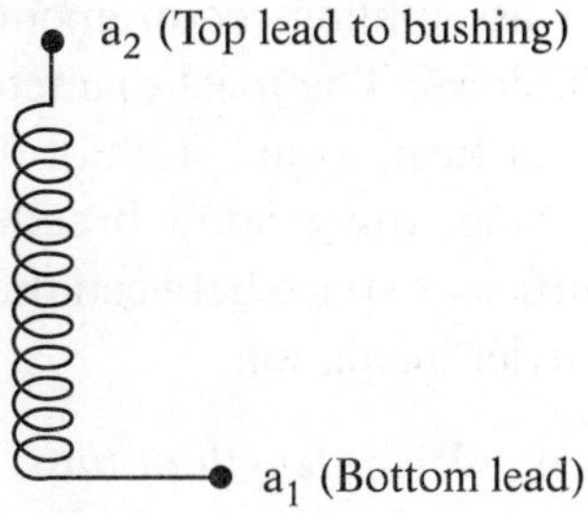

Fig. 1.23 *LV coil A phase limb*

Example 3: A representation of a 3 phase-transformer phases at the bushing terminals (Δ/Y transformer) is shown in Fig. 1.24.

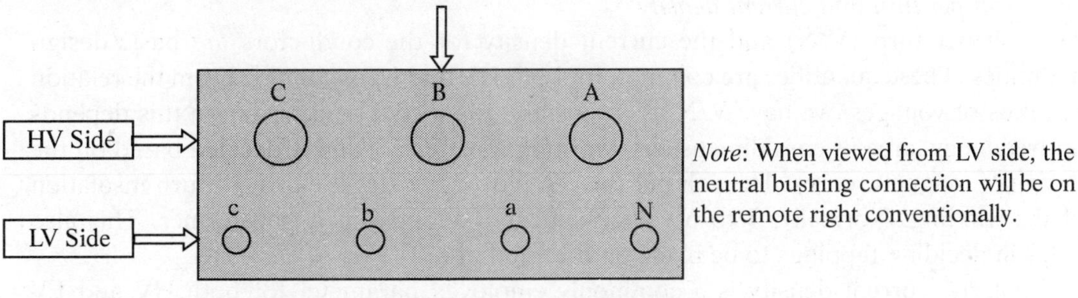

Fig. 1.24 *Top cover plan*

1.3.4 Design Consideration of Windings

(*i*) *Number of turns and size of wire*

The voltages of transformation decide the number of turns of the primary and secondary windings (N_P and N_S), whereas the capacity of the transformer decides the currents to be handled for the particular voltages and hence determines the size of wires to be used. Thus HV winding will have higher number of turns with a smaller cross section of wire while the LV windings will have lesser number of turns with a wire of higher area of cross section in a step down transformer.

(*ii*) *Size of wire based on temperature rise*
The capacity of transformer and the associated design of the size of wire also depend upon the effect of cooling. Hence, provision of proper ducts to carry away the heat in the core and windings (effective cooling system) can increase the factor of safety and lower the temperature rise.

(*iii*) *Core window dimensions*
The number of turns along with the size of wire and insulation decides the coil height and hence the window height of the core. This also includes the coil packing thickness for assembling requirement. The outer diameter of HV coil decides the breadth of core windows. The axial diameter of the outer winding is decided by the insulation used for inner turn, axial length of the layer, insulation cylinders, provision of oil ducts, etc.

Since the window breadth will be utilised by both the phase windings, there should be sufficient space between the two phase windings in the window for insertion of the phase barrier insulation.

(*iv*) *Mean length of turn*
The length of a turn is the circumscribed length of winding from the starting point to the completion of one turn, corresponding to the starting point. Depending upon the axial length of layer, this length of turn will vary. Hence, mean length of turn for the coil is considered.

(*v*) *Volt per turn and current density*
The volt per turn (V/N) and the current density for the conductors are basic design quantities. These quantities are common for both HV and LV windings. From the relation of ratios of voltages, we have V/N as a constant quantity. The decision on this depends on frequency, area of core dimensions, working flux density and is decided based on the design of magnetic cores. The volt per turn employed decides the turn-to-turn insulation of the winding. Normally, 3 to 6 volts per turn is adopted in a transformer. This also helps in deciding tappings to be made on the windings.

Similarly, current density is a commonly employed parameter for both HV and LV windings. A density of 2 to 4 A/mm^2 is adopted.

1.3.5 Winding Direction

The direction in which the winding turns are wound is very important. There are two winding directions, viz. the left-hand winding (LHW) and the right-hand winding (RHW). A left-hand winding is one in which the turns are wound in counterclockwise direction, while in right-hand winding, the turns are wound in clockwise direction.

The start of the winding viewed from the top fixes the direction, i.e., clockwise or anticlockwise. This is shown in Fig.1.25(a).

The various types of coils wound in LHW and RHW are shown in Fig. 1.25(b).

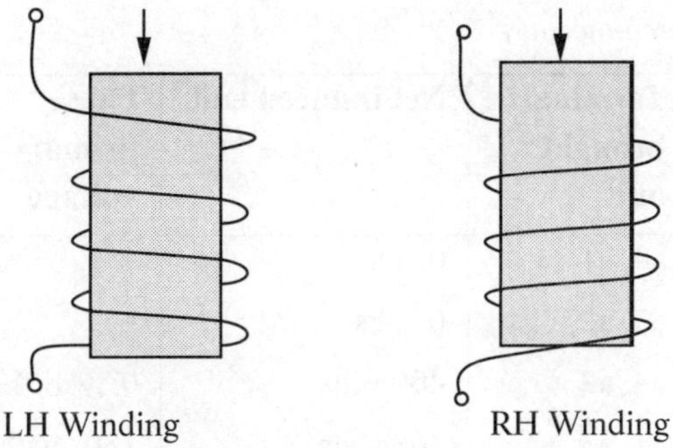

Fig. 1.25(a) *Winding direction*

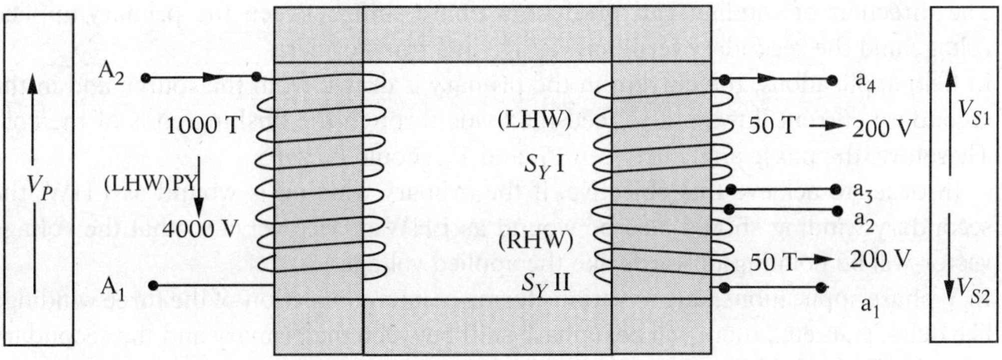

Fig. 1.25(b) *Types of coils wound in LHW and RHW*

The direction of the winding decides the direction of the induced emf. The induced emf in a left hand winding will be 180° opposite to that of the emf in a right hand winding wound on the same core. These emf can be added or subtracted, depending upon which terminals are linked and which terminals are taken out for connecting load. Utilising these terminals for interconnection also depends upon how the direction of windings is. This is elaborated in the following example.

Consider a core with primary and secondary windings. The secondary winding is shown in Fig. 124(b) in two equal parts: one with a RHW and the other with a LHW. The complete data of the turns, induced voltages, etc., are indicated. The emf are shown as vectors with the arrowheads pointing in the direction of the emf. Thus, V_P = 4000 volts is applied on the primary, and the secondary voltages available in the two windings as 200 volts each, i.e., V_{S1} and V_{S2} in opposite directions as shown in the figure.

If these two coils are to be connected in series to get a combined output, the following methods of interconnections will bring the results as indicated in Table 1.2.

Table 1.2 *Methods of interconnection*

Sl No.	Terminals interconnected	Terminals brought out	Net induced emf	Phase angle with primary applied voltage
1	a2, a3	a4, a1	0 volts	—
2	a1, a4	a3, a2	0 volts	—
3	a1, a3	a4, a2	400 volts	0° with V_P
4	a4, a2	a1, a3	400 volts	180° with V_P

The inferences from the above example are:

1. The direction of winding can produce a phase shift between the primary applied voltage and the secondary terminal voltage in a transformer.
2. In field applications, the current in the primary is drawn from the source and in the secondary current flows out to feed the loads through the bushing ends of the coil. Therefore, the phase angle between V_P and V_S should be zero.

   In order to achieve this objective, if the primary winding is wound as LHW, the secondary winding should also be wound as LHW or vice versa so that the voltage vector will be pointing upwards like the applied voltage V_P.
3. In 3-phase applications, due to various forms of interconnection of the three windings like delta, star, etc., there will be a phase shift between the primary and the secondary voltages. We call this a vector group.

   Hence, if the winding directions in the primary and the secondary are scrupulously followed according to the convention mentioned in point (2), it will not affect the phase shift due to the vector group.
4. The directions of windings should be noted before rewinding and repairs. If directions are changed, the net voltage will be reduced.

1.3.6 Types of Windings

The following types of windings are generally employed in power transformers.

1. Spiral type or cylindrical winding
2. Cross-over type
3. Helical type
4. Continuous disc type

The conductors used for the above windings are either (a) round section wires, (b) square section strips and (c) rectangular section strips (Fig. 1.26). Round wires are generally used with enamel insulation for current ranges of 15 A to 20 A only, i.e., 14 SWG or above. Square and rectangular section wires have paper covered or cotton covered insulations and are used for high current ranges.

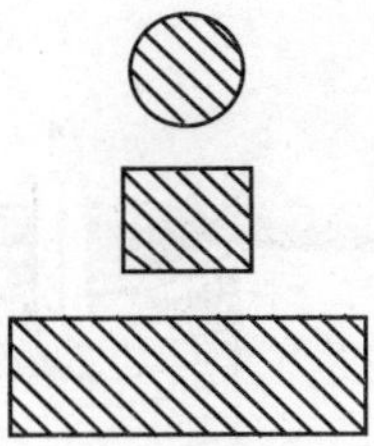

Fig. 1.26

A description of various types of coils used is given below.

Cylindrical winding or spiral type

Cylindrical coils are only suitable for carrying heavy currents and are hence used in LV side windings. These coils are wound on a "former" or directly over one insulating cylinder, which then will be used as insulation between the winding and the core limb. This may be a single layer or double layer type. Conductors of rectangular section when used are single or multiple conductors in parallel, continuously wound from top to bottom of the coil in the form of a spiral. Due to the width of the conductor, the turns will have a pronounced rake. Sufficient length of the conductor is left at each end for the lead connections, and the ends are bent to form these leads.

Edge block packing consisting of tapered strips of insulator is wound into the ends of the coil in order to form the coil as a true cylinder and to give it mechanical support when assembled on the core limb. This edge block is a split ring cut from a paper base laminate cylinder or coiled from a wedge made of several layers of pressboard taped together. To make the winding mechanically stronger, the edge blocks, together with the extreme turns of coil, are wrapped with tape wound half lap.

The single layer winding has its start and finish on opposite ends. In a double layer winding, the start and finish are arranged on the top of the coil and that the conductor passes from one layer to the other at the bottom. Here, the hand of the winding changes from layer to layer to accommodate the current direction reversal. There is one insulation cylinder between the two layers, and ventilating ducts are formed by inserting press board strips between the layers. The cylindrical windings of a single layer and double layer are shown in Fig. 1.27.

Note: The winding directions in these types will not change when the top and bottom ends are interchanged.

The diagram of a multiple-layer cylindrical winding with small size strip conductor and the details of layers and insulation, etc., is shown in Fig. 1.28.

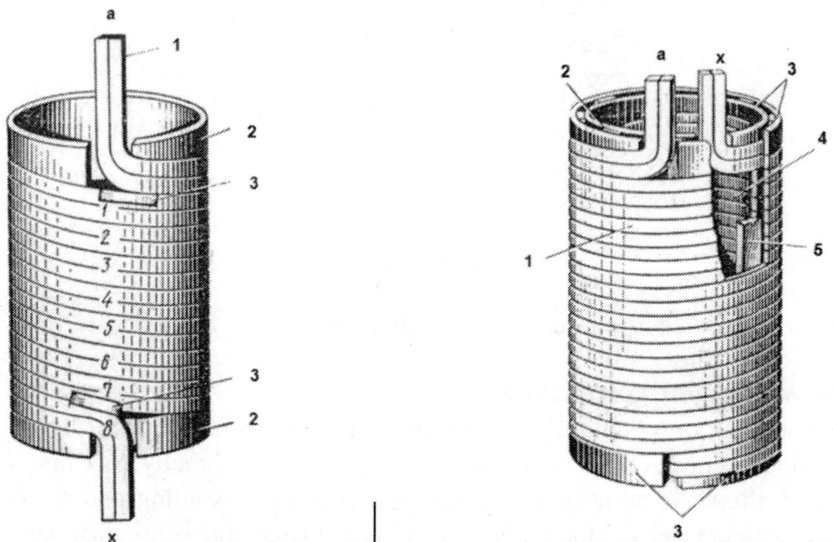

Single layer cylindrical winding wound with two parallel conductors
Number of turns - 8 (1 to 8)

a = Start of winding
x = Finish of winding
1 = Conductor for turns (2 strips in parallel touching each other)
2 = Edge blocks
3 = 'U' shaped boxes placed at the end where the conductor is bent (To improve inter-turn insulation at bending locations)

Left hand winding

Double layer cylindrical winding wound with two parallel conductors

a = Start of winding
x = Finish of winding
1 = First layer of turn
4 = Second layer of turn
2 = Vertical cooling ducts with strips
3 = Edge blocks
5 = Wood or pressboard slicks or strips for cooling

First layer = Right hand winding
Second layer = Left hand winding

Fig. 1.27 *Cylindrical windings of single layer and double layer*

Cross-over type winding

This is suitable for currents up to about 20 A and is largely used in HV winding of distribution transformers. These are wound on *formers*. Each coil consists of a number of layers with each layer having a certain number of turns. The conductors are generally round enamel wire or strip insulated with paper covering.

On completion of each layer, interlayer insulation paper is provided. This insulation is wrapped around the end turn of that layer and the turns for the next layer continues. This enables the whole coil to be kept intact and compact.

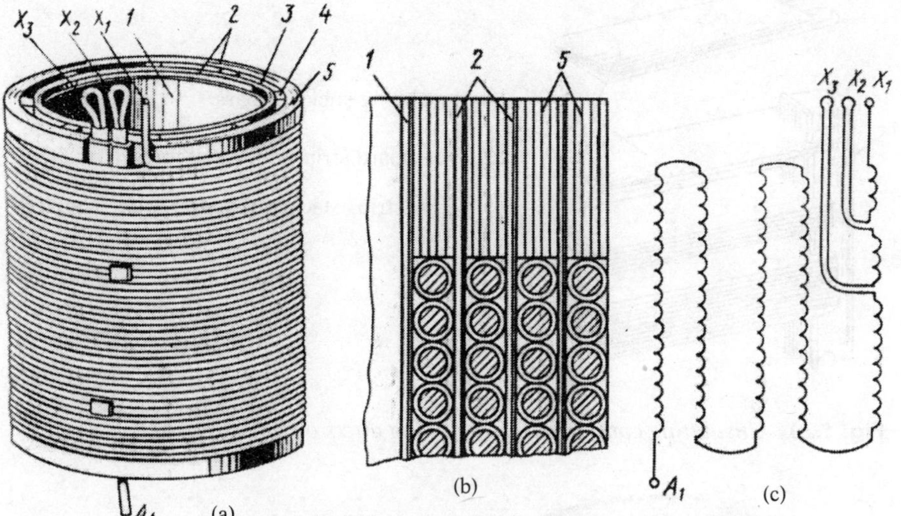

Multiple layer cylindrical winding

X1, X2, X3	Voltage control taps (FINISH)
A	Starting end
1	Paper base laminate cylinder
2	Inter layer insulation paper cylinder
3	Vertical ducts formed by sticks
4	Vertical bars or sticks
5	Press board strips inserted between interlayer insulation beyond coil layers

Fig. 1.28 *Multiple layer cylindrical winding*

Many such cross-over coils are provided in a phase winding and are connected in series. These coils are separated and spaced by spacers called *insulating pay sectors* and are dovetailed in the insulation strip kept around the limb portion over the LV coil. This is shown in Fig. 1.29.

The dovetail form permits assembly of spacers only by insertion from the top. Hence, these cannot be removed radially and remains mechanically strong in the event of forces arising due to short circuit, thereby keeping the coil position intact. The spacers also provide oil circulation facilities in between two cross-over coils placed on the limb. Figure 1.30 shows the cross-over coil of a transformer.

When a cross-over coil is placed in a limb, one end will be at the front top and the other end will be at the inner bottom. In order to avoid internal joints, this inside lead is heavily insulated and brought out through the gap formed between two coils and in between two sets of spacers. The cross-over coils are also called stacks when assembled over a limb (Fig. 1.31). In these cross-over coils, the hand of the winding must be carefully observed, since this will change when packed upside down.

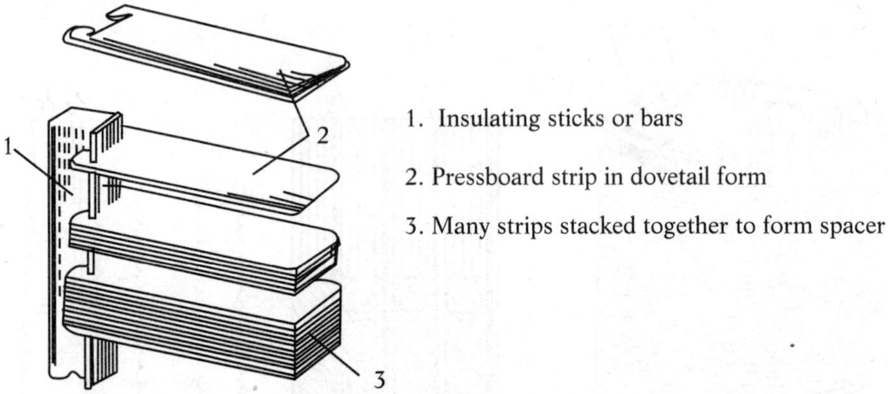

Fig. 1.29 *Insulating components for forming ducts on cross-over coil assembly*

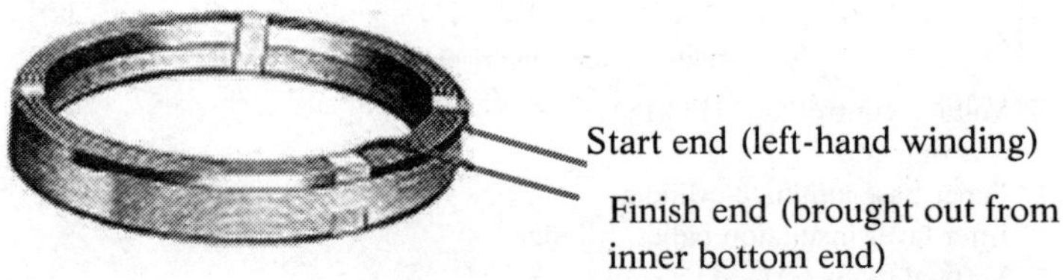

Fig. 1.30 *Cross-over coil (stack)*

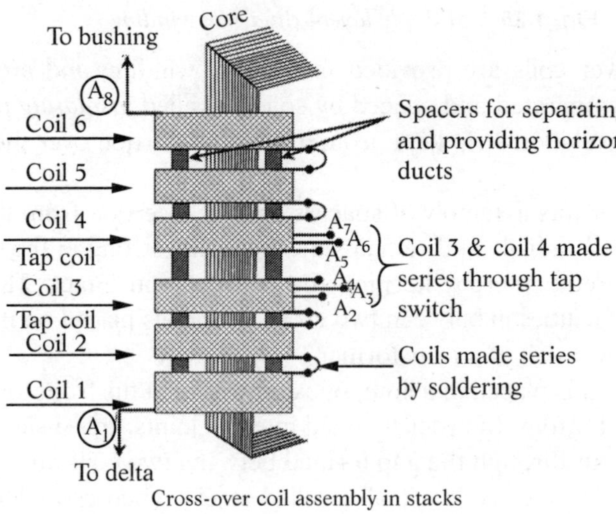

Fig. 1.31 *Cross-over coil assembly in stacks*

Helical type winding

In this the coil is wound in the form of a helix. It consists of a number of rectangular strip conductors wound in parallel, radially. Hence, each separate turn occupies the total radial depth of the winding. This is used for intermediate range of current and has comparatively smaller number of turns. The figure shows a helical coil wound with 8 parallel conductors per turn. These eight conductors are found spread radially on the radial depth and all these eight conductors in parallel forms a turn.

Sticks (or bars) are driven between the parallel conductors from the top to bottom and this forms the vertical oil duct. Radial key sectors (spacers) are provided between one turn and another, and this forms the radial oil duct. Hence, a system of oil ducts is provided in horizontal and vertical directions. Therefore, each separate conductor of each turn is in contact with oil.

The parallel conductors are arranged concentrically and are at different distances from the winding axis. Hence, conductors closer to the axis will have shorter length than those situated far away. Hence, transpositions are introduced at appropriate intervals to ensure uniform current distribution. Such transposition may be a group transposition or general transposition of all the conductors.

Figure 1.32 shows a helical winding with eight parallel conductors with transpositions, insulating cylinder and key spacers.

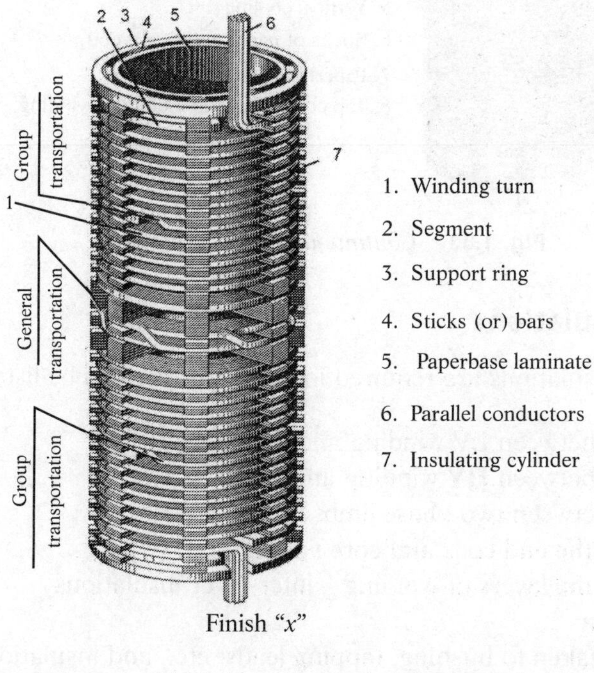

Fig. 1.32 *Helical winding with eight parallel conductors*

Continuous disc winding

This consists of a number of discs wound continuously with a single wire or with a number of strips in parallel. Each disc consists of a number of turns wound radially, one over another. On completing the radial end, the conductor uninterruptedly passes from one disc to the next disc. Hence, a full length of conductor without joints, as will be necessary for a complete winding, will generally be used when multiple strip conductors are used; transpositions are made at intervals to ensure uniform current distribution.

The discs are wound on an insulating cylinder, spaced from it by sticks running the whole length of the cylinder. The discs are separated by hard pressboard key sectors keyed to the vertical sticks. These horizontal and vertical spacers provide a system of ducts for oil circulation. The whole structure of the coils is mechanically sound and rigid and is capable of resisting stresses under most strenuous short circuit conditions.

The turn-to-turn insulation in a disc is however made of paper or cotton insulation strip only. Figure 1.33 shows a single conductor continuous disc winding.

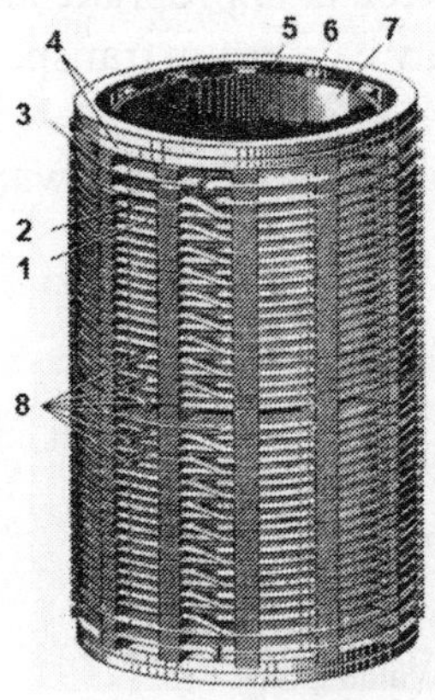

1. Disc or flat coil
2. Horizontal cooling duct
3. Pressboard spacer
4. Support insulation ring
5. Vertical cooling duct
6. Sticks of paper board or wood
7. Paperbase laminate cylinder
8. Tap connections in middle of winding

Fig. 1.33 *Continuous disk winding*

1.3.7 Winding Insulations

The places where the insulations are required in a transformer can be listed as follows:

1. Insulating cylinder between LV winding and the core
2. Insulating cylinder between HV winding and LV winding
3. Insulating barrier between two phase limbs
4. Insulation between the end coils and core yokes
5. Insulation between the layers of winding – inter-layer insulations
6. Inter-turn insulation
7. Insulation of leads taken to bushing, tapping leads, etc., and insulation of core bolts, vertical tie rods etc.(miscellaneous insulations)

Some of the insulating materials used for the above in a transformer are listed below.

(a) Pressboard sheets
(b) Elephantoid sheets
(c) Letheroid paper
(d) Hylum sheets
(e) Fibre tubes and washers
(f) Kraft paper
(g) Butter paper
(h) Enamel
(i) Varnish and shellac
(j) Cotton and yellow tapes
(k) Insulation sleeves of different sizes made of fibres

In the insulation requirements of a transformer listed above, the first two are special cylinders to fulfil the insulation requirement; they are made mechanically strong with a number of layers of specially selected pressboard or synthetic resin bonded cylinders. These cylinders are manufactured by winding paper coated with varnish on a *mandrel*, which is maintained at a fairly high temperature. The phase barrier is made of letheroid, pressboard or hylum sheets. Insulation between the end coils and core yokes consists of a series of pressboard washers and spacing blocks, or is built of insulating cylinders having internal and external diameter corresponding to the coil to be supported. The continuity of oil duct is also maintained by using suitable end washers.

The inter-layer insulation is made of Kraft paper or letheroid paper of suitable thickness, or butter paper, while the inter-turn insulation is made of enamel, paper or cotton covered on the conductor itself. Insulation of leads comprise suitably sized sleeves or fibre tubes fixed in treated wood or pressboard blocks through cleats. Similarly, fibre tubes or letheroid tubes are used for insertion of core bolts, tie rods, etc., and fibre washers are used for tightening the nut with core, yoke channel, etc.

Soldered or braced leads are covered with cotton or yellow tape, suitably impregnated with varnish and dried. The various *classes* of insulation and the details of insulating materials which come under these *classes* along with their maximum permissible hot spot temperatures are given in Table 1.3.

Normally, Class A insulation is adopted for distribution transformers, whereas Class B insulation is employed for power transformers of voltage class 110 kV and above.

1.4 Winding Coil Interconnections

In a 3-phase transformer, all the three phase windings are to be interconnected. This is done in three ways:

1. Delta

2. Star
3. Zigzag or interconnected star

The discussion in this section deals with the possible interconnections in vogue and the effects of such interconnections. The connections of coils adopted in double ratio transformers are also discussed.

Table 1.3 *Classification of insulating materials*

Sl No.	Class	Description	Max. hot spot temp.
1	Class Y	Cotton, silk, paper and paper products, pressboard, vulcanised fibre, wood, cellulose, fibre, PVC, VIR, etc., without impregnating substance	90° C
2	Class A	Cotton, silk, cellulose fibre, paper press board, vulcanised fibre, wood, etc., when impregnated or immersed in a dielectric like oil. Also cellulose acetate film, certain types of wire enamel, etc.	105° C
3	Class E	Superior wire enamels based on polyvinyl ferrul or epoxy resins, moldings with cellulose fibres, cotton fabric and paper laminates, epoxy resins, etc.	120° C
4	Class B	Glass fibre, asbestos, mica, built-up mica, glass fibre and mica laminates, etc., with bonding substances like shellac, bituminous compounds, epoxy resins	130° C
5	Class F	Glass fibre, asbestos, mica, varnished glass fibre textile, built up mica, etc. with suitable bonding substances like epoxy and silicon alkyd resins	155°C
6	Class H	Same as Class F but using appropriate silicon resins for bonding or impregnation	180° C
7	Class C	Mica, porcelain, ceramics, glass quartz, asbestos, treated glass fibre or treated asbestos with inorganic binders such as glass or cement, and silicon resins having superior thermal stability	Above 180°C

1.4.1 Connections in Practice

In our power system usage, whatever be the type of transformer (step up or step down), the primary winding is always connected in delta and the secondary in star to suit conditions of operation in India. No zigzag connections are employed except in some old distribution transformers.

1.4.2 Delta Connection in the Primary

The phase windings of the primary in a 3-phase transformer are connected in delta (see Fig.1.34), as seen from the HV (primary) side. The name delta refers to an equilateral

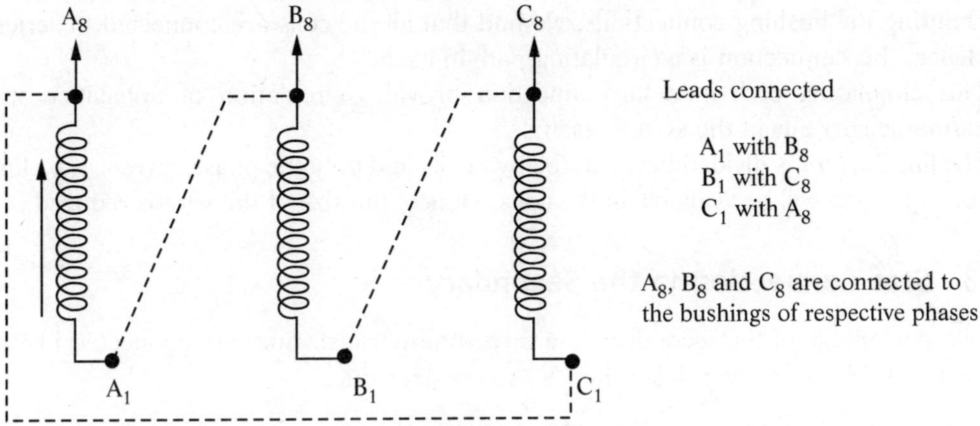

Fig. 1.34

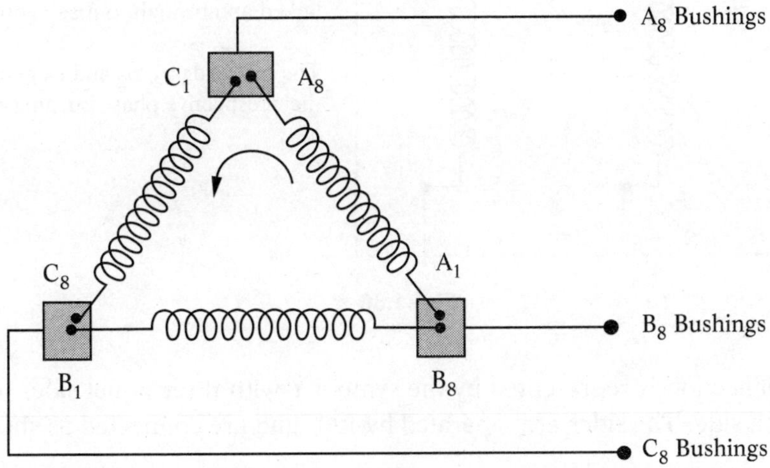

Fig. 1.35

triangle or Δ shape. The three-phase supply being a balanced one with a phase shift of 120°, this interconnection is called a delta connection (Fig. 1.35).

The characteristics of delta interconnections are as follows.

(a) Each phase coil is connected between two lines and therefore line voltage is applied across the coil. For example, if it is a 110 kV / 11 kV transformer, 110 kV is applied on the phase coil. Hence, the number of turns to be provided is according to this voltage only and the insulation to that level.

(b) The two ends of each phase coil are kept connected with the two line conductors.

(c) Since the ends of the coils are at line potential, no insulation grading is possible.

(d) Omitting the bushing connections, we find that all the coils are connected in series. Hence, this connection is a circulating path in itself.
(e) This circulating path in delta connection provides circulation of unbalance and harmonic currents in the system itself.
(f) The line current is divided between the two coils and only the phase current (i.e., line current $\div\sqrt{3}$) will be available in the coils. Hence, the size of the wire is reduced.

1.4.3 Star Connection in the Secondary

The phase windings of the secondary in a three-phase transformer are connected in star as depicted in Fig. 1.36, seen from the LV (secondary) side.

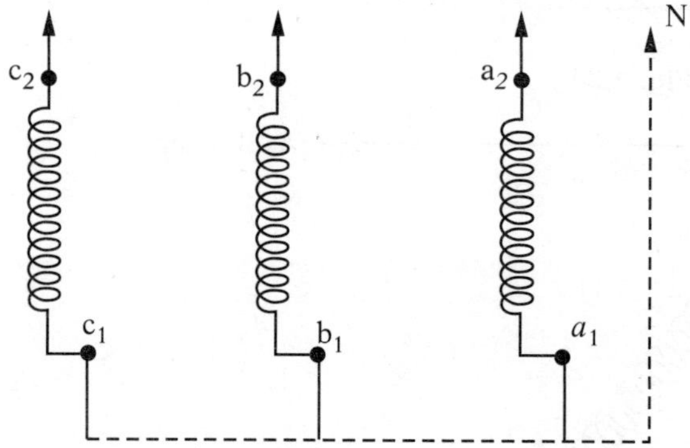

The bottom leads a_1, b_1 and c_1 are linked and brought out as neutral N.

The top leads a_2, b_2 and c_2 goes to their respective phase bushings.

Fig. 1.36

The star connection is represented by the symbol Y with three equal sides linked with one end of each side. The sides are separated by 120° and are connected as shown in Fig. 1.37.

The characteristics of star connection are listed below.

(a) The ends of each phase coil are connected to the lines. Hence, only phase voltage is available in each coil. For example, in a 110 kV/11 kV transformer, 11 kV appears on the secondary side between the two lines, whereas the induced voltage in each coil is 11kV $\div\sqrt{3}$ volts. Hence number of turns in the secondary phase would be in conformity with phase voltage only.
(b) In a three-phase circuit, if one of the terminals of all the three coils is linked, the net potential according to the three-phase theory is zero and hence we get a *neutral* for earthing purposes.
(c) Since one end of the coil starts at zero voltage, insulation grading is possible.
(d) The line current flows directly through the phase coil and hence the size of wire is to be selected accordingly.
(e) This type of connection does not have a circulation path in itself.

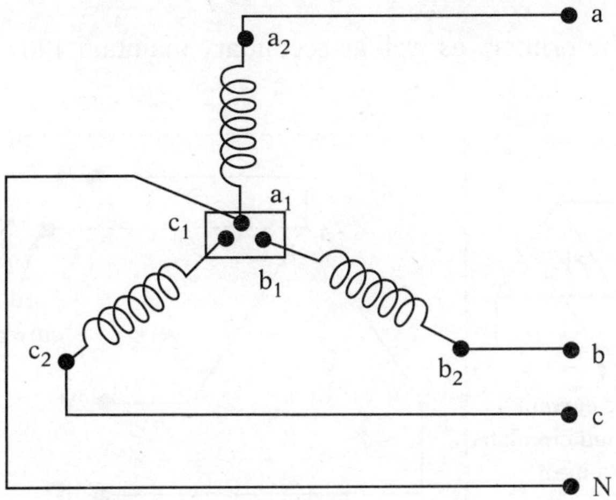

Fig. 1.37

1.4.4 Characteristics of Delta/Star Combination and Vector Group

A combination of delta in the primary and star in the secondary of a transformer enables the following, e.g., in a 110 kV substation power transformer:

(a) A star point for solidly grounding the voltage level at the sending end.
(b) An unbalanced four-wire supply system may be adopted so that balanced and unbalanced loads are supplied simultaneously.
(c) Third harmonic voltages are eliminated by the circulation of 3rd harmonic current on the primary side.
(d) In case of earth fault on the secondary side, the earth fault current (zero sequence current) flowing through the earth will enter through the grounded neutral to the secondary winding. This is possible because the primary is in delta (Fig. 1.38). This unbalance will be reflected in EHV side and gets circulated in delta itself without appearing on the EHV line. Hence, earth fault relay on the secondary side alone will sense and act. The E/L relay on the EHV side of a substation or at the EHV sending end will not act. Thus, a definite jurisdiction for the earth fault relay to act is established.

Vector groups

Each phase undergoes a phase shift when passing through a Δ/Y transformer. The phase shift is between

(i) R phase primary and R phase secondary
(ii) Y phase primary and Y phase secondary

(iii) B phase primary and B phase secondary

while R, Y and B of the primary as well as secondary maintain 120° phase shift among themselves.

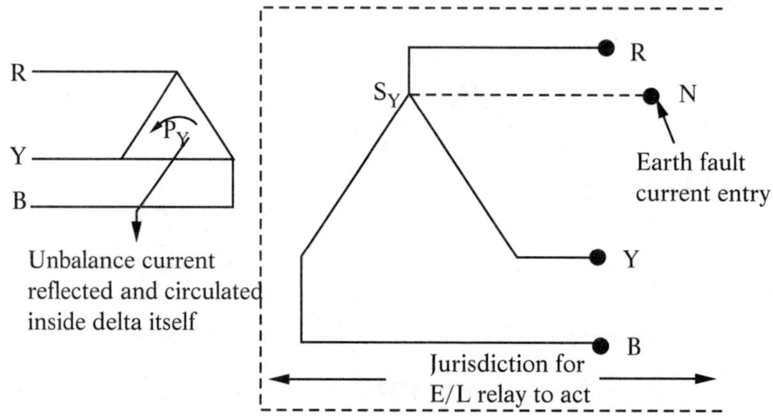

Fig. 1.38

Different combinations of connections between the primary and secondary of a transformer produces different phase shift and these combinations are defined as vector groups. The delta–star combination which has been discussed so far comes under vector group 4_1DY11. In this, the various symbols are:

4_1	–	Group number and sub group number
D	–	Connections in the primary winding
Y	–	Connections in the secondary winding
11	–	Phase shift between phases of primary and secondary denoted by the 11 o'clock position on the clock face.

The 11 o'clock position represents +30° phase shift, i.e., the angle between the hour hand and minute hand at 11 o'clock.

Keeping the delta vectors of the primary side (supply side) as reference and drawing the corresponding parallel star side vectors of the secondary winding, we find that the star is entirely tilted by +30° (anticlockwise) corresponding to the reference star in the primary as shown in the dotted lines of Fig 1.39.

Other vector groups

Each vector group represents an angle of shift obtained and a subgroup represents the connections by which this angle of shift is obtained. Usually, three combinations (subgroups) are available in each group. Four such main vector groups are classified as below:

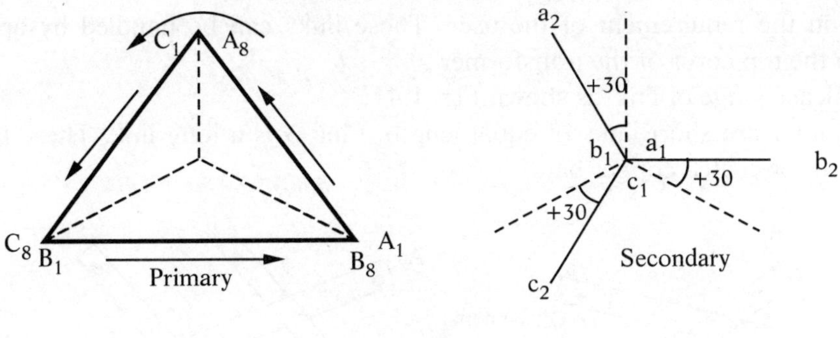

Fig. 1.39

Group 1	0°	1_1YY0 1_2Dd0 1_3Dz0
Group 2	180°	2_1YY6 2_2Dd6 2_3Dz6

Group 3	−30°	3_1DY1 3_2Yd1 3_3Yz1
Group 4	+30°	4_1DY11 4_2Yd11 4_3Yz11

1.4.5 Interchangeable Vector Group for 4_1DY11

Though 4_1DY11 transformers are used in general in India,, sometimes 3_1DY1 transformers are also used since the two are interchangeable. The delta interconnection made in the opposite direction is shown in Fig. 1.40.

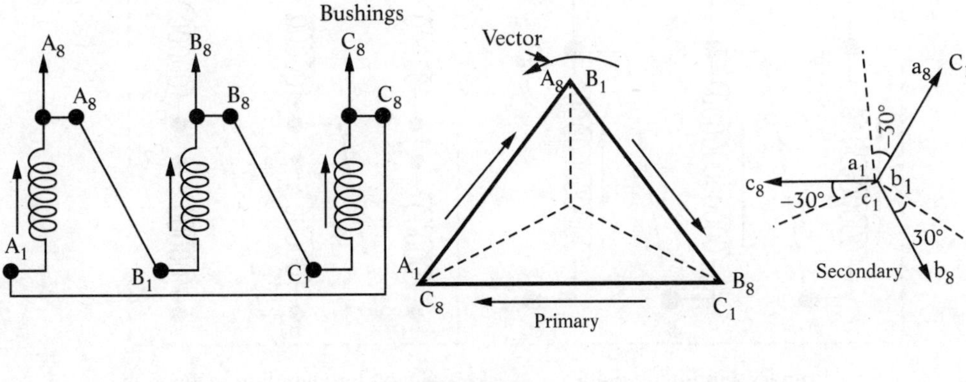

Fig. 1.40

In order to facilitate this interchanging, the terminals of coils $A_1B_1C_1$ and $A_8B_8C_8$ are brought to the top of the transformer. The required delta is made through links. By interchanging these links, either 3_1DY1 or 4_1DY11 vector group can be obtained depending on the requirement of the user. These links can be handled by opening a manhole on the top cover of the transformer.

The physical change of links is shown Fig. 1.41.

Links 1 and 2 are short links of equal length. Link 3 is a long link. These links are made up of silver coated copper bars.

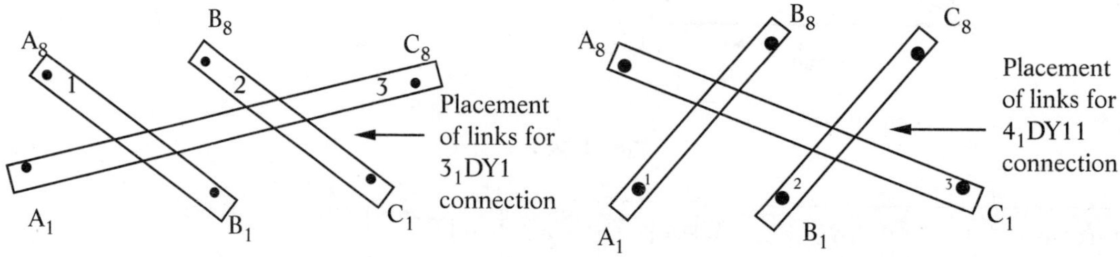

Fig. 1.41

1.4.6 Zigzag or Interconnected Star

In some old distribution transformers there may be zigzag connection in the secondary with delta on the primary. This has all the characteristics of a delta/star connection but there is no phase shift between the primary and secondary. Each secondary phase winding is split into two as shown in Fig. 1.42 to enable zigzag connection. The figure gives a 1_3Dz0 connection and the corresponding vectors.

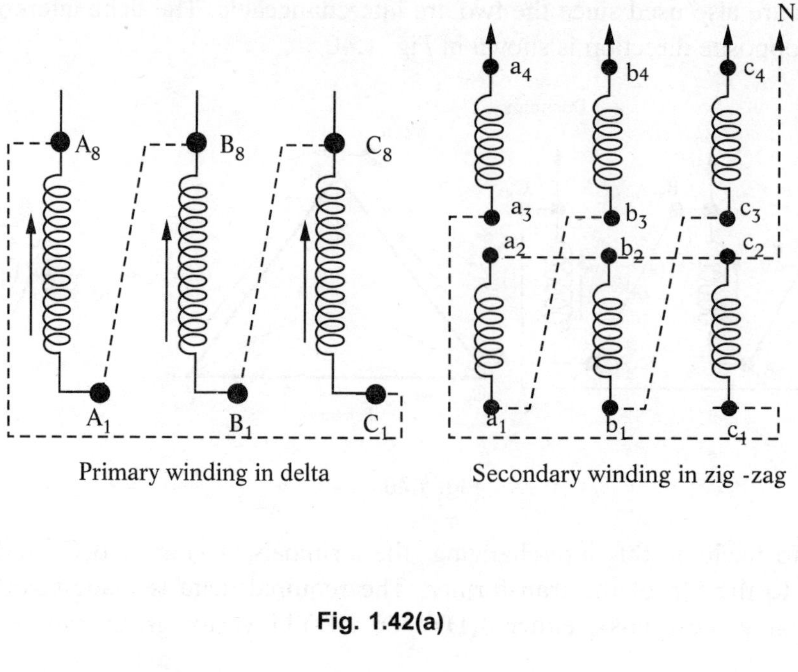

Fig. 1.42(a)

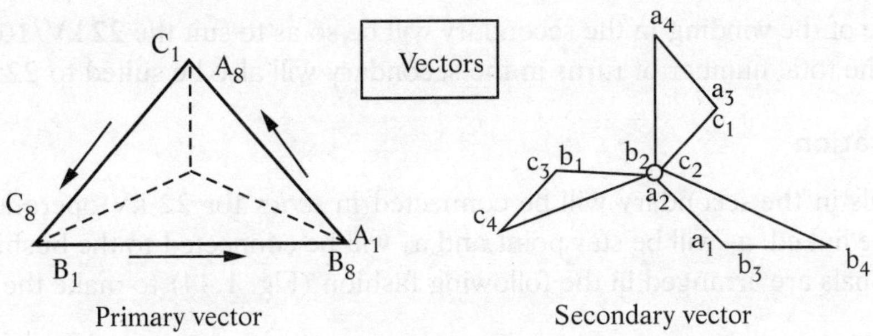

Fig. 1.42(b)

1.4.7 Coil Connection for Double Ratio Transformer

In employing power transformer in a sub-station, it may become necessary to use a double ratio transformer so that different voltage levels can be employed in the secondary. The connection of coil terminals for these different usages is discussed here.

For example, a 10 MVA delta / star transformer has a voltage ratio of 110 kV / 22-11 kV. So it has the facilities to set the secondary voltage either at 22 kV or at 11 kV. The secondary coils is split into two halves and the terminals are brought up to the top of the transformer so that setting up of the required ratio can easily be done by opening a manhole on the top cover and making connections through links as in the case of vector group setting.

The connections of one of the secondary phase coil for each voltage setting are shown in Fig. 1.43

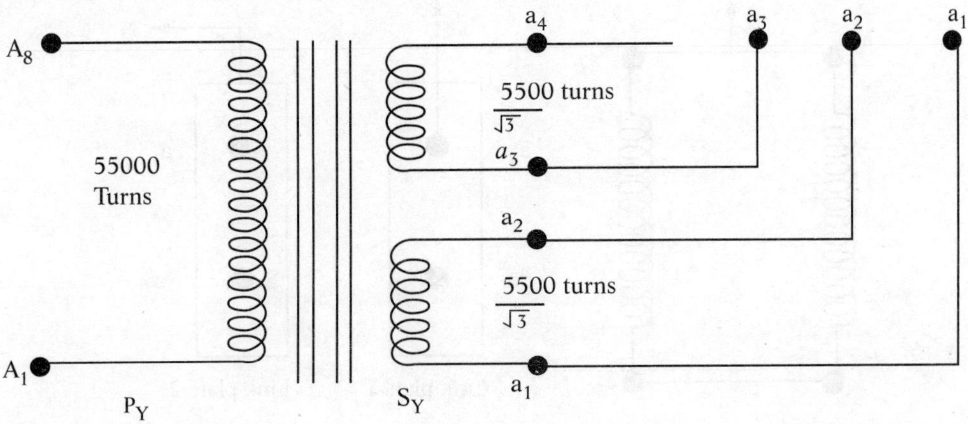

Fig. 1.43

Considering 2 volts/turn, we have 55000 turns in the primary and $11000/\sqrt{3}$ turns in secondary. The secondary turns are divided into two coils with $5500/\sqrt{3}$ turns each and terminals a_1, a_2, a_3 and a_4 are brought out as shown in Fig. 1.43.

The gauge of the winding in the secondary will be so as to suit the 22 kV/10 MVA cap design and the total number of turns in the secondary will also be suited to 22 kV.

22 kV operation

The split coils in the secondary will be connected in *series* for 22 kV operation, i.e., a_2 and a_3 will be linked. a_1 will be star point and a_4 will be connected to the bushing.

The terminals are arranged in the following fashion (Fig. 1.44) to make the link.

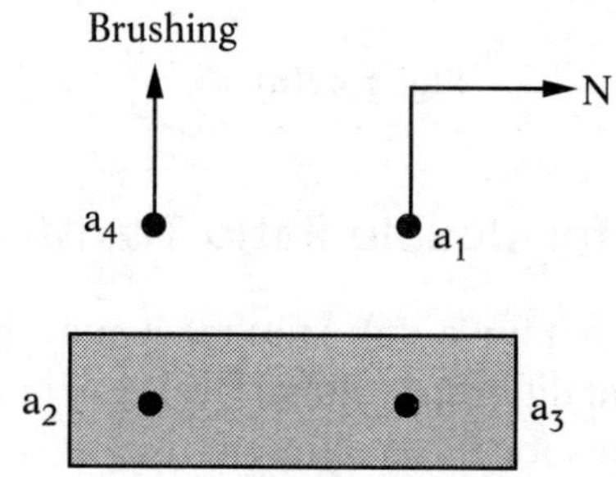

Fig. 1.44

11 kV operation

For this, the two coils in the secondary are connected in parallel and the links made as shown in Fig. 1.45.

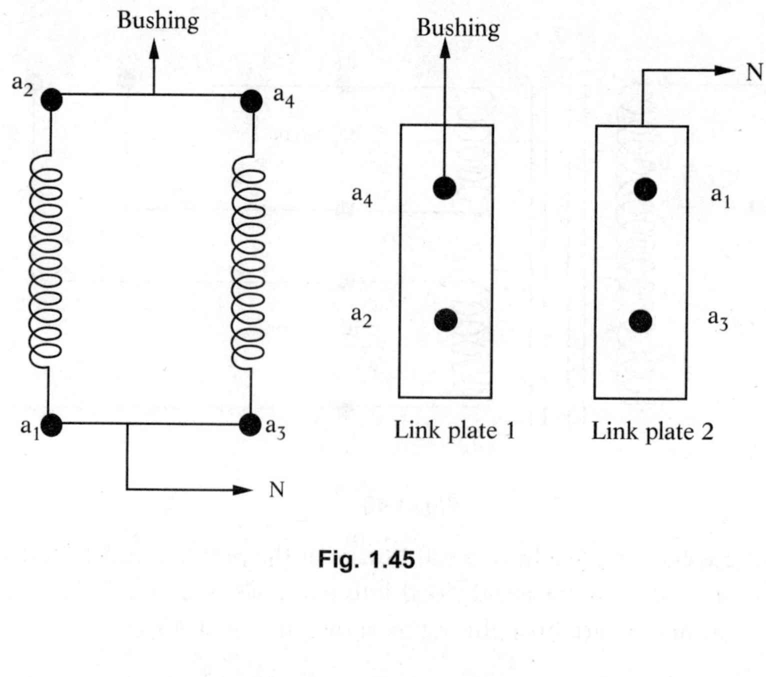

Fig. 1.45

Usually, there are two link plates. For 22 kV operation, both link plates are combinedly placed, connecting a_2 and a_3. For 11 kV, individual links are used for connecting a_2 - a_4 and a_3-a_1.

The coils are arranged for 11 kV operation as mentioned above because the complete coils spread over the entire height of the winding is used so that heat dissipation is uniform and cooling system is effective. Secondly, the winding wound with the gauge for 22 kV operation will be used for 11 kV operation also by paralleling the split windings for to accommodating the higher current in the 11 kV operation.

1.5 Tappings and Tap Changers

One of the simplest and most inexpensive methods of providing adjustments in supply voltages is to arrange tappings on transformer windings. The purpose of such arrangement in a transmission and distribution system is to maintain the secondary voltage constant with varying primary voltages. This section deals with the method of tapping the windings, connection to tap switches, voltage adjustments obtained, on-load tap operation, etc.

1.5.1 Voltage Control by Tappings

The reliability of a power system is judged by its ability to maintain uninterrupted supply to loads, whereas the quality of the system is judged by its ability to maintain the rated frequency and voltage within tolerable limits at the terminals of the consumers. The frequency limits are maintained by the generators of the power system only, whereas the control of voltage can be made by the following means.

(a) Adjusting the excitation of generators to supply required VARs.
(b) Employing synchronous condensers, capacitor banks, etc., and thus supplying the required VARs.

The above two methods are effective in transmission systems where the VAR flow is considerable and the voltages at various points depend on these VARs only. Hence, control of VARs will improve the voltage condition in the transmission system. Conversely, raising of the voltage by the use of tappings in transformers will control VARs to a certain extent.

But in distribution systems, the voltage variation may not be fully attributed to VARs. The following methods are also employed effectively for voltage controls in distribution systems.

(c) Installing booster transformers where broad limits of voltage regulation is required.
(d) Adopting tappings in transformers.

Therefore, for transformers installed at substations, it is easier to arrange tappings to control the voltage.

Use of tappings for voltage variation

Voltage control by tappings is based on the principle of changing the turns ratio, i.e., by changing the active turns in one of the transformer windings. The control can be on the increasing trend or the decreasing trend.

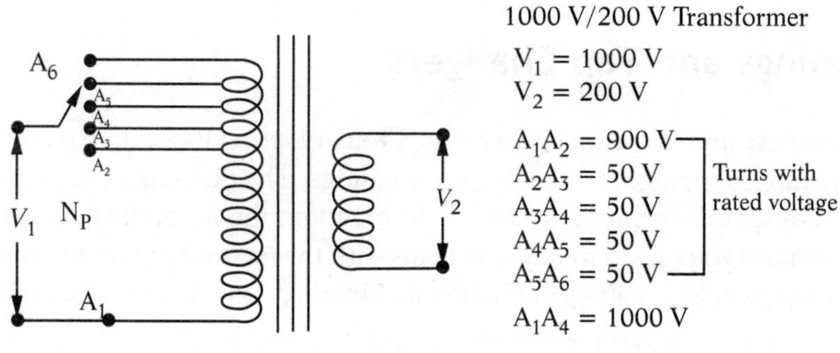

Fig. 1.46

For example, consider the primary and secondary windings shown in Fig. 1.46, where the requirement is to keep V_2 constant. Therefore, depending upon the available voltage V_1 on the primary side, the active turns on the primary winding are to be changed. The description of tappings is shown. The moving arm can choose any one of the tappings, i.e., A_2, A_3, A_4, A_5 or A_6.

In the normal case, the arm will be at A_4. Thus, A_1A_4 will be the active winding ready to receive the normal voltage of 1000 volts, so that 200 volts are maintained at the secondary. The turns ratio is $1000 \div 200 = 5$.

If the incoming voltage is only 950 volts, then the arm is set at A_3, thereby decreasing the turns by $5\%(-5\%)$ from the normal. Now, A_1A_3 is the active winding ready to receive the available 950 volts in the primary side, and the 200 volts in secondary is maintained constant by reducing the turns ratio to $950 \div 200 = 4.75$.

Similarly, if the incoming voltage is 1050 volts, i.e., 50 volts above the normal, then the arm is set at A_5 by increasing the turns by $5\%(+5\%$ of normal) from the normal. Now, A_1A_5 will be the active windings ready to receive the incoming 1050 volts in the primary side and the 200 volts in the secondary is maintained constant by increasing the turns ratio to $1050 \div 200 = 5.25$.

If the number of turns in the primary is reduced, then it is a decreasing trend of turns referred to as the *minus percentage of normal*, and if it is increasing, then it is called an increasing trend of turns referred to as the *plus percentage of normal*.

It is customarily referred to as only the decrease or increase of turns in percentage of tappings and not the voltage increase or decrease in the secondary for maintaining constant magnitude on the secondary side.

1.5.2 Method of Tappings

It is preferable to take tappings from the high voltage side because of the following reasons.

(a) It is possible to arrange the tappings within very fine limits to give any desired voltage.
(b) The tappings on the LV side will have severe limitations with regard to the percentage of tappings. The number of turns in the LV will be low, which limits the fineness of adjustment. Also, it is the high current side.
(c) If tappings are made in the LV windings, the tapping leads have to be brought out carefully, crossing the HV winding.

The tapping leads from the winding is made in three forms. One is by making the tappings in the upper end of the winding to which the bushings are connected. The second is to make it in the lower end of the winding to which the delta formation is made. Thirdly, tappings can be arranged in the middle of the winding. Of the three methods, the first two are employed in on-load tap changers and the third is used generally in off-load tap changers. The tappings made in the middle is more preferable since it is in the electrical centre of the winding, and hence the magnetic equilibrium between the primary and the secondary winding is less disturbed and the axial forces acting between them are thereby reduced, providing greater mechanical stability of the windings.

At present, a separate tap winding is provided in the middle for on-load tap changing with special operating arrangement. This is described at the end of this section.

Tappings in the upper end of the coil

Figure 1.47 shows the arrangement of tappings at the upper end of the coil. The lead is connected to the bushing and the delta point is connected to the moving arm selector which selects the taps.

This arrangement is shown for A phase winding. Similar arrangements are available for B and C phase windings.

Tappings in the lower end of the coil

This arrangement is shown in Fig. 1.48. The delta lead connecting the bottom end of the coil is connected to the moving arm selector contact that selects the tap. The tap selector arrangements are brought to the top of the transformer for easy operation from outside the tank.

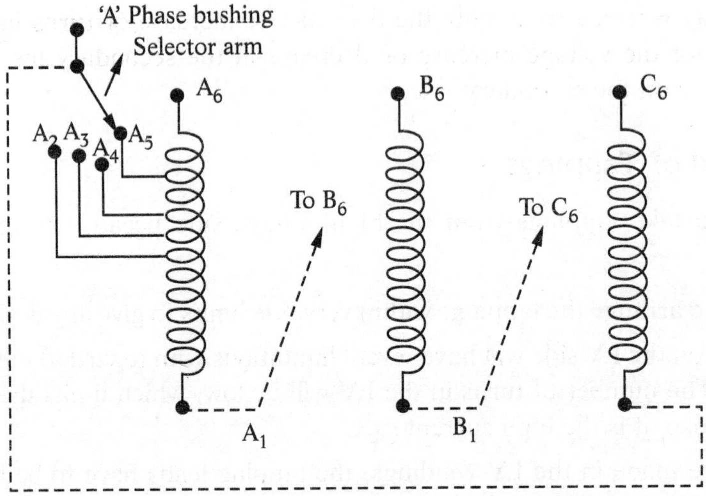

Fig. 1.47

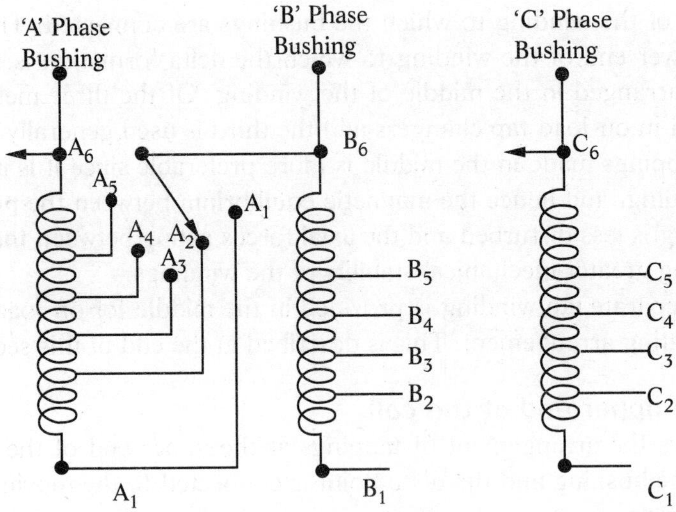

Fig. 1.48 *A phase winding*

Tapping in the middle of the coil

This is the usual method of tap changing in off-load tap changers and it is advantageous from the operational point of view. The arrangements are discussed below in detail.

The whole phase winding is split into two halves and tappings are made at the bottom of the top half-section and at the top of the bottom half section. Usually, these coils are crossover types in low capacity transformers. Hence, a number of such coils will be stacked to form the whole phase winding. The number of such stacks will be even so that it will be possible to divide it into two halves for arranging the tapping. For example, if 10 stacks are used in a phase winding, first five stacks will be for top half and the next five stacks are the bottom half. Then the fifth and sixth stacks will be tapping coils. The above tap arrangements in a phase winding are shown in Fig. 1.49.

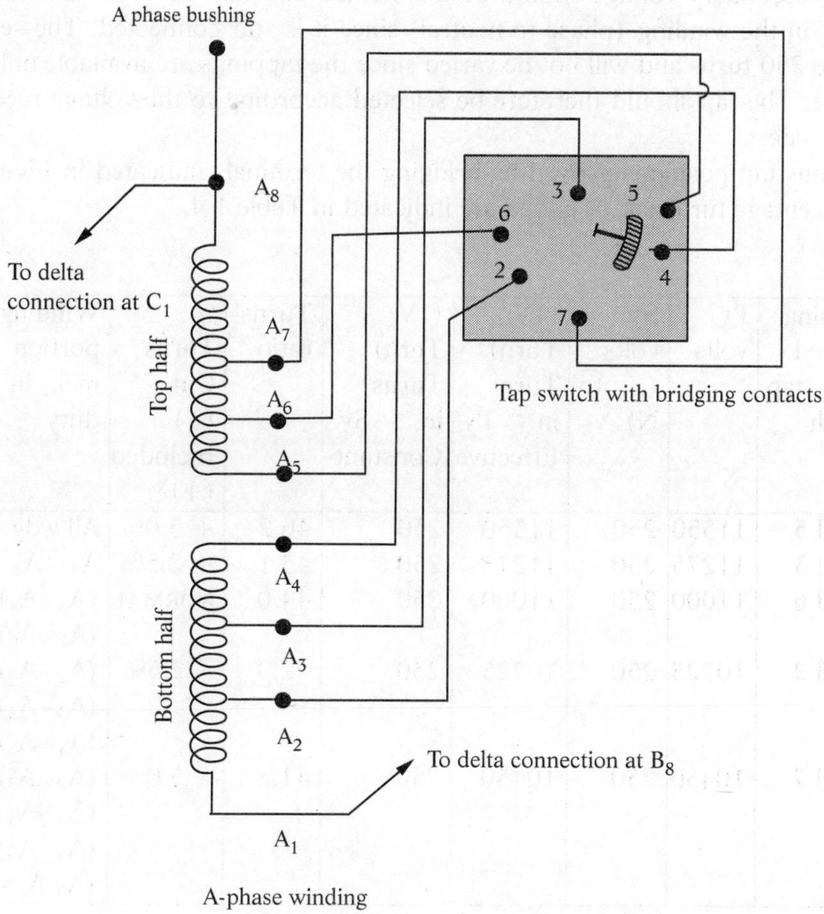

Fig. 1.49 *A-phase winding*

In this arrangement, the tapping leads A_2, A_3 and A_4 relate to the bottom half of the winding and the leads A_5, A_6 and A_7 relate to the top half of the winding. In the previous two methods, the selector selects only a particular tap by making contact with it. But here, one of the bottom half leads is joined with one of the top half leads to set the required tapping. Care should be taken to see that the leads of the same half are not bridged, as this will result in short circuit of the turns lying in between the two bridged leads. To ensure this, the terminal connection in the tap selector switch is so arranged alternatively in the manner indicated in the sketch. The selector contact bridges two terminals in a sequence and thus selects the tapping.

For example, consider a distribution transformer of an 11 kV / 433 volts (delta / star) with tappings +5% to −5%. Assuming a 1 V / turn design of the transformer, the primary will have 11000 turns in the normal position to receive the required primary voltage of 11 kV. The secondary voltage should be maintained constant at 433 volts line-to-line or at 250 V in the winding (phase to neutral) since it is star connected. The secondary turns will be 250 turns and will not be varied since the tappings are available only on the primary side. The tap should therefore be selected according to the voltage received on the primary side.

The various tap positions arrived by bridging the terminals indicated in Figure 1.49, and the percentage turns cut or added are indicated in Table 1.4.

Table 1.4

Tap No.	Terminals bridged by tap switch	Py. volts	Sy. volts (φ to N)	(1V/ Turn) Turns in Py Effective	(1V/ Turn) Turns in Sy. Constant	Turns ratio	% Turns Cut (−) / Included (+)	Winding portion not incl. in active duty
1	4 and 5	11550	250	11550	250	46.2	+ 5.0%	All wdg. incl.
2	5 and 3	11275	250	11275	250	45.1	+ 2.5%	$A_3 - A_4$
3	3 and 6	11000	250	11000	250	44.0	NORMAL	(A_3-A_4) and (A_5-A_6)
4	6 and 2	10725	250	10725	250	42.9	− 2.5%	(A_2-A_3) and (A_3-A_4) and (A_5-A_6)
5	2 and 7	10450	250	10450	250	41.8	− 5.0%	(A_2-A_3) and (A_3-A_4) & (A_5-A_6) & $(A_6\text{-}A_7)$

Note: There is another bridging between 4 and 7 in the rotation of tap selector. This will give the normal position as in Tap 3 by deleting two tap section windings (A_5-A_6) and (A_6-A_7) from active duty. But this is not generally used since deletion of windings is not symmetrical.

The power transformer in a substation will generally have a tapping facility of +3% to −9%. For this, the terminal connections must be so arranged in the tap switch so as to have the sequence of required variations.

1.5.3 Effect of Tapping on Impedance of Winding

The reactance of the winding is susceptible to changes due to the presence of tappings. The reactance is proportional to the number of turns and its mean length and inversely proportional to volt/turn. The cutting out/in of appreciable turns for the voltage variations by tappings tends to distort the leakage field and hence the reactance values change. Such changes will be appreciable when the tappings are made at the end of the coil (lower or upper) and will be low when it is made at the electrical centre of the winding. These aspects of variations of reactances due to tappings must be carefully considered in the parallel operation of transformers.

1.5.4 Off-load Tap Changers

The selection of tap setting after the supply to the transformer is switched off is called *off-load tap changing*. This is more advantageous because tap connections can be changed without load current and proper setting of selectors on the contacts can be safely ensured before switching on the supply to the transformer.

The tapping lead inside the tank is connected to the selector placed internally, operated by an external handle from outside the tank. This selector is for a three phase operation. The contacts for the three phases will be arranged independently, but is simultaneously operated through a gang operated common shaft so that simultaneous change of taps are made for all the three phases.

These off-load tap changers may be any one of the forms mentioned below.

(a) Drum type changers with slip ring contacts
(b) Drum type changers with segmental contacts
(c) Rack type changers with knife blade contacts

The changers are made of an insulating body (such as Hylum) with sufficient thickness to bear the load due to tap change operation and has contact terminals placed at appropriate places. The common insulated handle of the shaft is projected from the tap switch. This handle is connected with a metal rod through bolts and nuts or cotter pins and brought out externally for operation. Leak proof arrangements are made at the tank where the handle comes out by placing gaskets and asbestos rope.

Drum type changers with slip ring contacts

The tap leads are connected to the terminals on the one side and the corresponding contacts on the other side. The common shaft has a slip ring firmly fixed through spring. This spring ensures low contact resistance of the slip ring with contacts and also ensures that the ring bridges two terminal contacts and does not rest in the middle. The arrangement is shown Fig. 1.50.

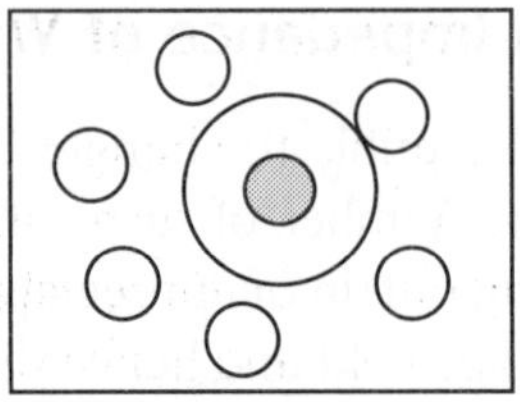

Fig. 1.50

This type of tap changers are placed horizontally on the top of core in a metal frame, and all connections are brought to the terminals as shown Figure 1.51.

Fig. 1.51 *Core-coil unit of transformer with H V off-load tap selector*

Drum type with segmental contacts

This type of tap changer has a vertical arrangement placed on the face of the core and is properly sealed between the top and the bottom yoke channels with leads brought to the terminal. The insulating handle is connected with a metal rod and brought to the top cover of the tank with oil leak proof arrangements and tap operation is done externally at the top of the tank. Figure 1.52 shows these arrangements. The segmental contacts is housed in a fibre tube in a common long shaft as shown in the figure.

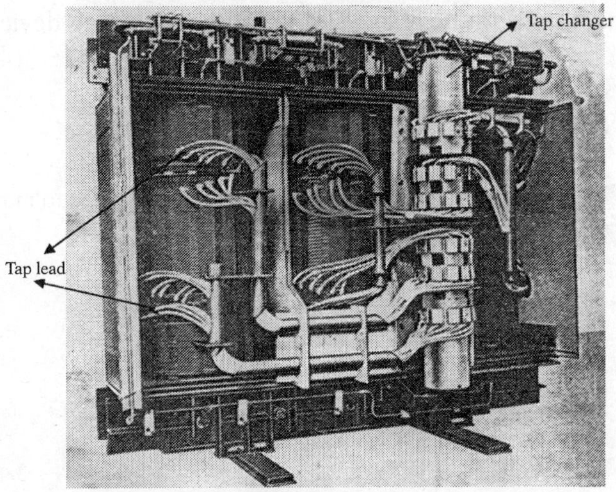

Fig. 1.52

Rack type changers with knife blade contacts

This type of changer is housed in the top of the transformer core. A common rack with three knife blade contacts for the three phases moves over the fixed brass contacts and can be operated externally from the side of the tank or the top cover. This arrangement is shown Figure 1.53.

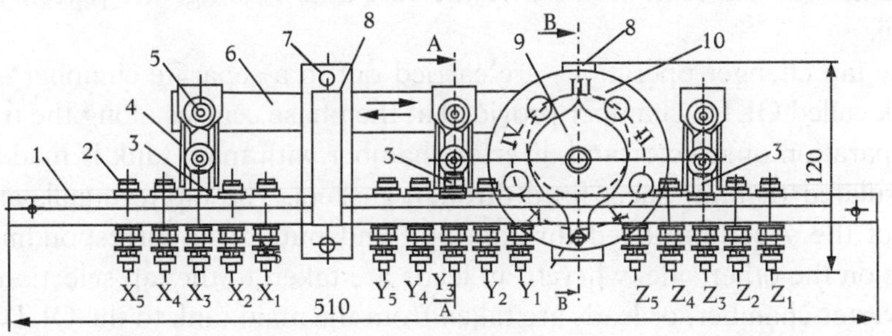

1 Common frame to house contacts
2 Fixed brass contacts to which tap leads connected
3 Movable contact
4 Coil springs
5 Fixing bolt
6 Common rack moving system
8 Glass-fabric base laminated material holding contact frame (fixedly) and moving rack
7 Fixing of frame 8
9 Operating knob
10 Tap indicating dial

Fig. 1.53 *Rack type changer*

The knife-blade contact used in the above system of tap changing device is shown in the Figure 1.54.

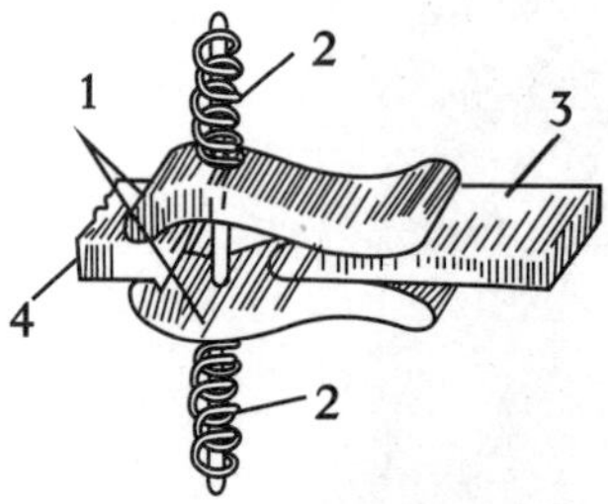

1 Curved copper strips (JAWS)

2 One end of the Jaws attached to moving system

3 Pressing springs

4 Fixed blade (Tap terminals)

Fig. 1.54 *Rack type changer knife blade contact*

1.5.5 On-load Tap Changers (OLTC)

The change of tap settings under varying load conditions without interruption of supply is called *on-load tap changing*. The main difficulty here is that since the contacts are carrying the load current, change-over must be done smoothly with minimum arcing and carbonisation. This is ensured if quick and smooth transition is made from one position to the other. Since the change-over causes arcing and carbonisation, the operation cannot be performed in the main tank itself where the core and windings are placed, as this will pollute the oil.

Hence, the tap changer operations are carried out in a separate chamber annexed to the main tank called OLTC chamber provided at the phase centres along the transformer tank. The separation of selector and diverter chamber with main tank is made with very thick and insulated (hylum) sheet fixed through gaskets. The tap terminals are brought to one side of the sheet and fixed through bolt and nuts. The corresponding through terminals are on the other side, wherefrom leads are taken to the tap selection terminals inside the diverter chamber, or leads are taken from the main tank to the OLTC chamber through oil leak proof glands. All tap change operations are carried out in this separate chamber only through mechanical arrangements with lever and gears linked with this chamber.

Arrangements in diverter chamber

The OLTC devices must have two main fundamental features, namely

- some form of impedance which is introduced to prevent short circuiting of the tapping section
- a duplicate circuit provided so that the load current can be carried by one circuit while switching is being undertaken on the other.

The impedance can take the form of either a resistor or a centre tapped reactor. Hence, on-load tap changers are classified as resistor type or reactor type.

In earlier days, the reactor type was more popular than the resistor type. This is because the reactor can be continuously rated whereas the resistor has a definite time rating due to high power dissipation service. But the introduction of high speed resistor transition tap changing has made the resistor type common in the present day. This is achieved by building the stored energy drive in the switching device itself. Also, the contacts are made with copper-tungsten alloy arcing tips, thereby improving the contact life.

The working of the above two types is discussed below in detail.

Reactor type

The centre tapped reactor type has two selectors, one for the main and the other for the duplicator circuit. The reactor is connected to the two selectors through two diverter switches. The tapping is made at the centre of the reactor to connect the load circuit. The arrangement is shown schematically in Fig. 1.55.

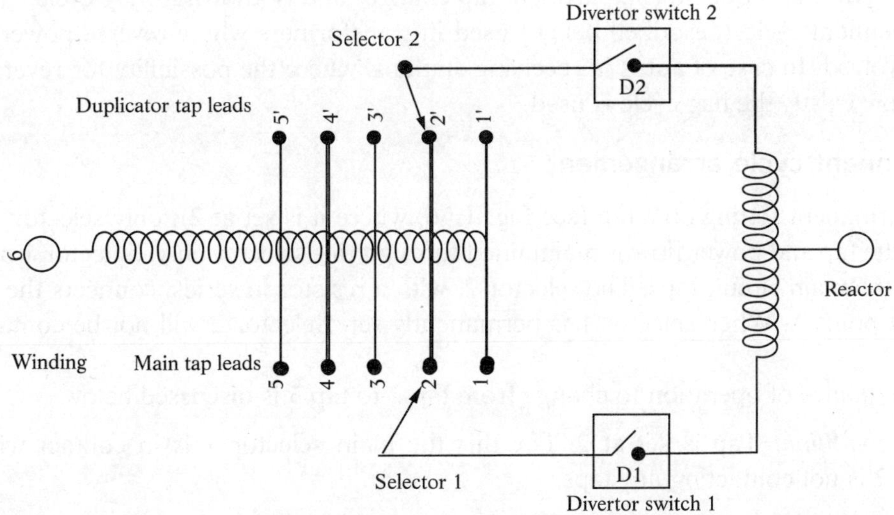

Fig. 1.55 *Reactor-type changers*

The sequence of operation for the tap change operation under load is given below.

Existing condition: Selectors 1 and 2 are at tap number 2, diverter switches 1 and 2 are kept closed.

1. Diverter switch 1 is opened.
2. Tap selector 1 moves to tap 3. During the above movement, the load is maintained through selector 2 contact, through diverter switch 2 and through half of the reactor winding.

3. Diverter switch 1 is re-closed. Now the winding between tap 2 and tap 3 is bridged by the reactor and a circulating current flows through the reactor.
4. Diverter switch 2 is opened.
5. Tap selector 2 is moved to tap 3′ from 2′. During this period, load is maintained through selector 1, diverter 1 and half of the reactor winding.
6. Diverter switch 2 is re-closed. Now the load is being maintained in parallel from tap 3 and tap 3′ through diverter switches 1 and 2 and through the reactor.

In this permanent operating position, the impedance drop across the reactor is eliminated since the current flow from each parallel path is in the opposite direction.

Resistor type

There are two types in this method of on-load tap changing. One is called single resistor type, which carries out selection and switching on the same contacts and use one transfer resistor. This involves a single compartment tap changer and is known as pennant cycle.

The other type has separate tapping selectors and separate diverter switches using two resistors. This has a double compartment tap changer and is known as flag cycle.

The pennant cycle (described below) used in transformers where reverse power flow is not involved. In case of autos in receiving stations, where the possibility for reversal of power flow exists, the flag cycle is used.

(a) Pennant cycle arrangement

In the permanent setting on a tap (see Fig. 1.56, where it is set at 2), only selector 1 will contact the tap and power flow is maintained through this selector only. Selector 1 moves through the main circuit taps. The selector 2, with a resistor in series, connects the main circuit at point A. When selector 1 is permanently set, Selector 2 will not be contacting tap 2′.

The sequence of operation to change from tap 2 to tap 3 is discussed below.

Existing condition: Tap is set at 2. For this the main selector 1 is in contact with 2. Selector 2 is not contacting any taps.

To set the tap at 3.

1. Selector 2 moves and contacts tap 3′. The winding between tap 2 and tap 3 is bridged by the resistor and a circulating current flows through the resistor.
2. Selector 1 moves from tap 2. Now the transition resistor carries load current.
3. Selector 1 contacts tap 3. The load current flows in parallel in the two circuits, through selector 1 and selector 2.
4. Selector 2 is released from contact 3. Load current now flows through the main selector 1 only from tap number 3. The tap is now set at 3.

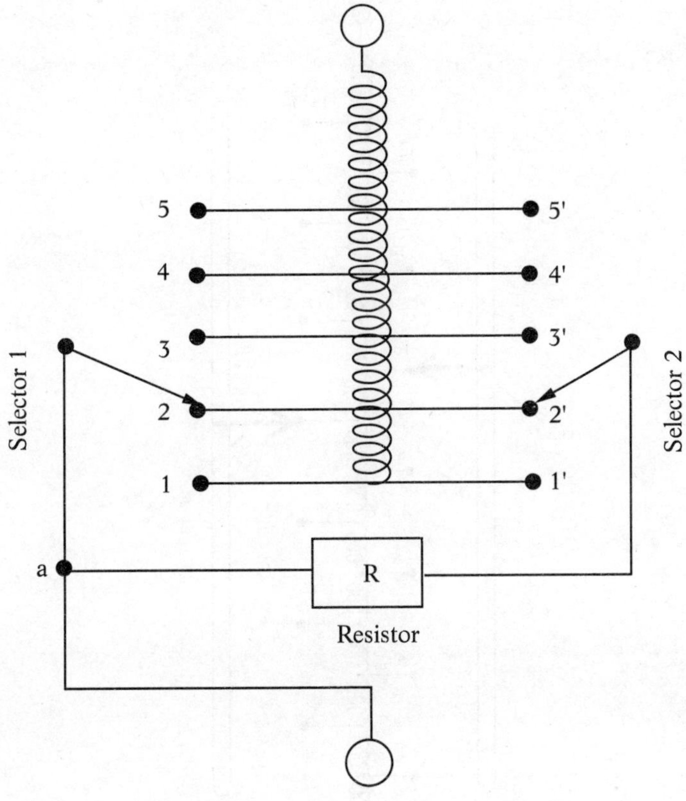

Fig. 1.56 *Resistor type changers*

(b) Flag cycle arrangement

This is the usual arrangement made in larger transformers (see Fig.1.57). The tap selectors are arranged generally in circular form. Linear selector arrangement (in-line form) are also be used, since this provides a greater voltage withstanding capability over the tapping range.

The diverter switches have contacts operating in rapid sequence with usually four separate make and break units. Here instead of having duplicator tap circuit, separate taps are maintained on either side for each selector. The typical selector and switching arrangements is shown in Fig.1.57. There are 14 taps and 15 positions. Two selectors, 1 and 2, for even tap selection exists, and odd tap selection exists separately. The taps are arranged to facilitate in-line movement of the selectors over crescent arrangements. Each selector circuit has diverter switches which have a main contact and an auxiliary contact referred to as (M_1, A_1) and (M_2, A_2) respectively in the figure. The auxiliary contact circuit has a transition resistor in series, shown as R_1 and R_2 respectively. The main bus bar of these switches is connected to the neutral point of the transformer winding.

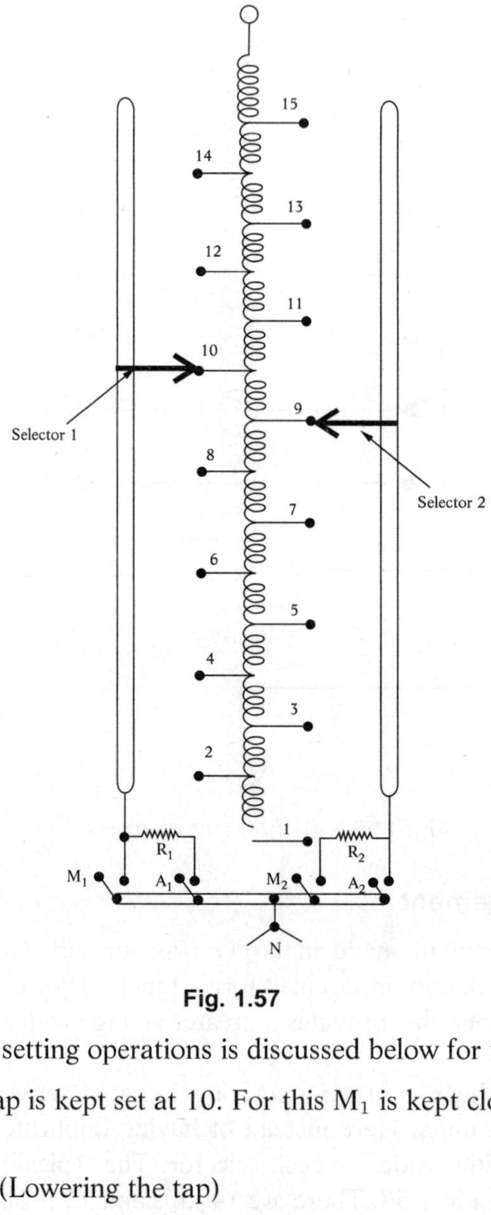

Fig. 1.57

The sequence of tap setting operations is discussed below for this arrangement.

Existing condition: Tap is kept set at 10. For this M_1 is kept closed, A_1, M_2 and A_2 are kept open.

Now to set the tab at 9 (Lowering the tap)

1. Close auxiliary contact A_1. Load current hitherto flowing from tap 10 through M_1 only, now takes a parallel path through R_1 and A_1 also.
2. Open main contact M_1. Load current from tap 10 flows through R_1 and A_1 only, which are in series.
3. Close auxiliary contact A_2. The two resistors R_1 and R_2 are in series across tap 9 and tap 10, set up a circulating current through them while maintaining the load current, which divides passing through each of the resistors to each of the tappings.

4. Open auxiliary contact A_1. The circulating current is now interrupted. The load is now transferred to tap 9, passing through resistor R_2 and contact A_2.
5. Close main contact M_2. Load current flows through tap 9 in a parallel paths, one through the M_2 circuit and other the through R_2 and A_2 circuits.
6. Open auxiliary contact A_2. Load current flows through tap 9 and through main contact M_2 only.

The tap is now set at 9.

If the tap change is to be done from tap number 9 to tap number 10, the sequence described above is to be reversed.

It may be noted that because odd and even tappings are arranged for the two selectors independently, for a reversal of the sequence, i.e., for lowering from increasing trend or for increasing from lowering trend, the selectors need not be moved. In the operating sequence described above, selector movement is not there because selector 1 is set at tap 10 and selector 2 is set at tap 9 already.

If we want to move the tap to tap 11 from tap 10, the selector should be moved from tap 9. The sequence is described below.

Existing condition: Tap is kept set at 10. For this M_1 is kept closed, A_1, M_2 and A_2 are kept open.

Now to set the tap at 11. (Increasing the tap)

1. Move selector 2 from tap 9 to tap 11.
2. Close auxiliary contact A_1.
3. Open main contact M_1.
4. Close auxiliary contact A_2.
5. Open auxiliary contact A_1.
6. Close main contact M_2.
7. Open auxiliary contact A_2.

The tap is now set at 11.

Note:

(1) All the above sequences are usually carried out in a time duration of 50 to 80 milliseconds.

(2) The diverter main and auxiliary switches opening/closing are arranged in the form of gliding contacts.

(c) Flag cycle arrangement with separate tapping coil in the middle

The present usage of on-load tap changer involves a separate tapping coil, placed in the middle of the phase winding. Moreover, 17 taps are being availed of from the tap coil with just 9 tapping leads (8 taps) only. This is achieved in two ways: (i) by permanent offset of voltages kept cut off through the tapping coil and reusing the tap coil, and (ii) by current reversal principle. For this arrangement, a pre-selector switch is also employed in addition to the selector and diverter switches.

(i) *Permanent offset arrangement*

This arrangement is shown in Fig. 1.58 as used in a 10 MVA power transformer of 110 kV / 11 kV with primary taps ranging from +5% to −15%, with each tap for 1.25, % i.e., 1375 volts.

The tapping coil wound for 11 kV with 9 leads brought out is shown in the middle, with lead numbers from 5 to 13. When all the 11 kV is cut off, it is offset permanently to a main coil having lead numbers 3 and 4. The lead numbers 3, 4 and 5 are kept connected to a pre-selector switch. The main winding is in two parts, one on top portion wound for 46.75 kV, plus the 11 kV offset with lead numbers 14 and 2 as shown in figure. Lead number 1 is connected to bushing and lead number 2 is connected to delta. Lead 14 is connected to the selector contacts.

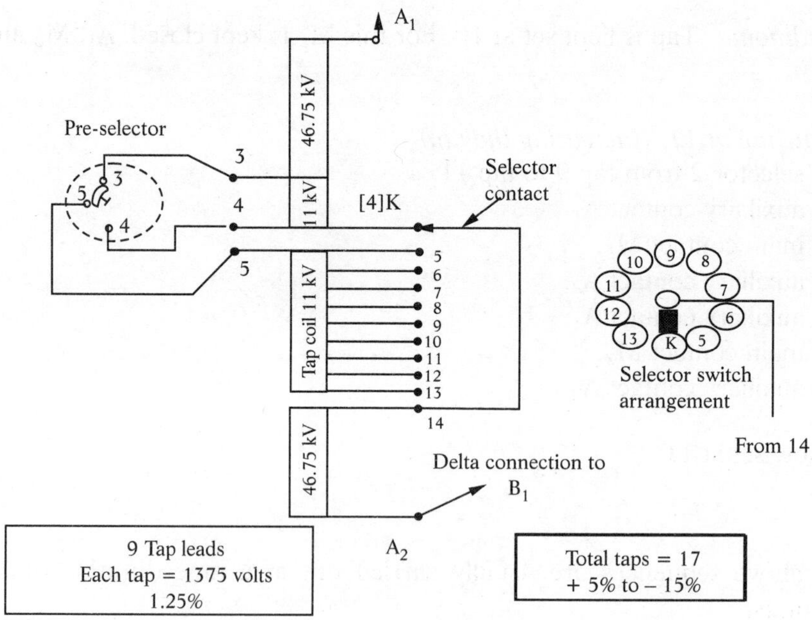

Fig. 1.58 *Tap arrangement - 10 MVA 110 kV / 11 kV transformer*

Table 1.5 shows all the taps with voltages on primary, current handled at each tap, secondary voltage, selector contact position and pre-selector position at each tap for lowering as well as raising operations.

It may be noted that the normal tap, i.e., 110 kV / 11 kV, occurs at tap 5. A transition position is catered to by a tap to 9 as 9A, 9B and 9C, with same voltage of 104.5 kV but with different positions of the selector contacts and pre-selector contacts. In this, tap 9 is catered to by 9B only. Hence, during lowering operation, i.e., from tap 10 to tap 9, the mechanism is so arranged as to skip off 9C and set at 9B only. Similarly, in the raising operation, i.e., from tap 8 to tap 9, tap 9A will be skipped and set at tap 9B only.

The transition stage is for permanent offset of the voltage kept cut off through taps. For example, during raising operation with the pre-selector between 4 and 5, the selector goes from lead 13 to lead 5 and is set at lead 4 (K), thus cutting off voltages up to 11 kV. This 11 kV is permanently offset by changing the pre-selector contacts position from 4 and 5 to 3 and 5. The selector contacts then moves to lead no. 13 and includes the tapping coil in the circuit and again proceeds to cut off for another 11 kV. Thus, 17 taps are achieved by just 8 taps.

Table 1.5

Tap No.	HV volts	HV amps	LV volts and amps	OLTC tap connection	Pre-sele. position Raise	Lower
1	115500	50.0		14 – 13	5.4	5.4
2	114125	50.6		14 – 12	5.4	5.4
3	112750	51.2		14 – 11	5.4	5.4
4	111375	51.8		14 – 10	5.4	5.4
5 [N]	110000	52.5		14 – 9	5.4	5.4
6	108625	53.1		14 – 8	5.4	5.4
7	107250	53.8		14 – 7	5.4	5.4
8	105875	54.5	11000 V 525 A	14 – 6	5.4	5.4
9A	104500	55.2		14 – 5	5.4	5.4
9B	104500	55.2		14 – 4 (K)	5.4	5.3
9C	104500	55.2		14 – 13	5.3	5.3
10	103125	56.0		14 – 12	5.3	5.3
11	101750	56.7		14 – 11	5.3	5.3
12	100375	57.5		14 – 10	5.3	5.3
13	99000	58.3		14 – 9	5.3	5.3
14	97625	59.1		14 – 8	5.3	5.3
15	96250	60.0		14 – 7	5.3	5.3
16	94875	60.8		14 – 6	5.3	5.3
17	93500	61.7		14 – 5	5.3	5.3

The on-load tap change operation scheme is defined in Fig. 1.59.

	Existing tap set at 14	→	*To be changed to 13*
1	Selector movement	:	Selector 1 already at lead number 8
		:	Selector 2 set at lead number 9
		:	Pre-selector set between 3 and 5

2	Diverter movement	:	Existing feed through M_1
	a) Bridges M_1 and A_1	→	Load current passes through M_1 and A_1
	b) Released from M_1	→	Load current passes through A_1 only
	c) Bridges A_1 and A_2	→	Load current passes through lead numbers 8 and 9. Tap voltage is taken care of by R_1 and R_2
	d) Released from A_1	→	Load current passes through A_2 only
	e) Bridges M_2 and A_2	→	Load current passes through M_2 and A_2
	f) Released from A_2 and settles at M_2	→	Tap set at 13

The odd and even tap leads are separately taken through two selector contacts, 1 and 2, and connected to the contacts in the diverter chamber. The diverter chamber consists of two main contacts, M_1 and M_2, each linked with an auxiliary contacts A_1 and A_2 through resistors R_1 and R_2. The tap change operation is carried out only in the diverter chamber. There is a gliding contact in the diverter chamber, which moves and bridges the main and auxiliary contacts sequentially.

Selector 1 is connected to M_1 and selector 2 is kept connected to M_2. The selector movement is only for selection of tap before tap changing. The movements for change of a tap is described in the figure. It may be seen that the two taps are bridged during the tap operation at one stage of diverter movement. While bridging A_1 and A_2, R_1+R_2 is presented as a load for the voltage between two taps and thus avoids short circuit, taking care of the circulating current.

The load currents are diverted through parallel bridging and is hence not interrupted. However, bridging of A_1 and A_2 will interrupt the circulating current due to the difference of voltage between the two taps. This may only cause slight carbonisation in the diverter chamber.

(ii) *By current reversal principle:*
This arrangement is used in a 100 MVA 230/110 kV auto transformer with primary tap ranging from +10% to −10%, with each tap set for 1.25% (i.e., $2875 \div \sqrt{3}$) volts.

The arrangement of separate tapping coil and main windings is shown in Fig. 1.60(refer to Table 1.6 also).

The arrangement is as described previously in the offset arrangement. But here, instead of voltage offset, the reversal of current through the tap coil is adopted.

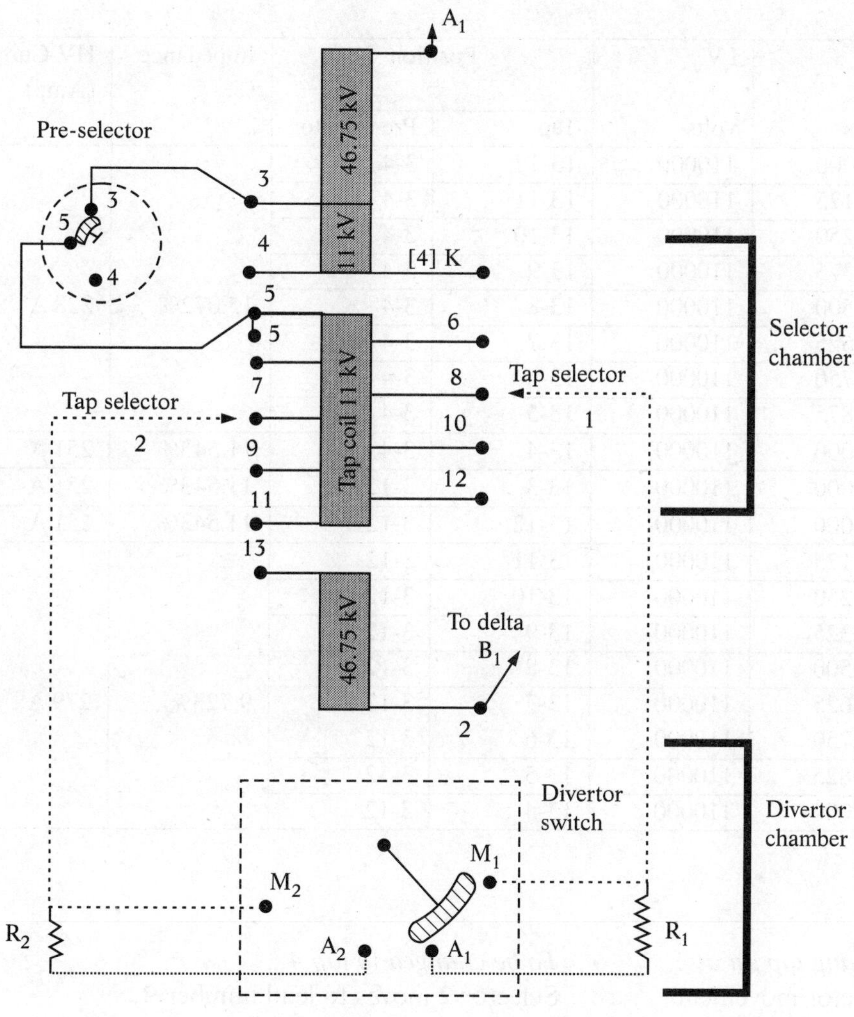

Fig. 1.59 *Tap operation – 10 MVA 110 kV / 11 kV Transformer*

Hence, the pre-selector switch is connected with terminal 3 from the bottom of top main winding and the terminals 4 and 12 of the tapping coil. When 3 and 4 are bridged and selector contact is in lead number 12, the same current flows in the tapping coil as that of the main winding. The transition stage at tap 9 (9A, 9B and 9C) will alter the bridging of pre-selector contacts from 3 and 4 to 3 and 12, reversing the current direction in tapping coil. Now, while increasing the taps, moving through tap leads by selector contact from 12 to 4 will add negative voltage and further reduce the voltage (Fig. 1.61).

All other operations for change of taps is similar to that described earlier. An example of tap changing with diverter, selector switches is described below for the tap operation scheme shown in Fig.1.61.

Table 1.6

Tap no.	HV volts	LV volts	Position		Impedance %	HV Current (Amp)
			Tap	Pre selector		
1	253000	110000	13-12	3-4		
2	250125	110000	13-11	3-4		
3	247250	110000	13-10	3-4		
4	244375	110000	13-9	3-4		
5	241500	110000	13-8	3-4	15.072%	228 A
6	238625	110000	13-7	3-4		
7	235750	110000	13-6	3-4		
8	232875	110000	13-5	3-4		
9A	230000	110000	13-4	3-4	11.543%	251 A
9B	230000	110000	13-3	3-12	11.543%	251 A
9C	230000	110000	13-12	3-12	11.543%	251 A
10	227125	110000	13-11	3-12		
11	224250	110000	13-10	3-12		
12	221325	110000	13-9	3-12		
13	218500	110000	13-8	3-12		
14	215625	110000	13-7	3-12	9.723%	279 A
15	212750	110000	13-6	3-12		
16	209825	110000	13-5	3-12		
17	207000	110000	13-4	3-12		

Existing tap set at 5 → *To be changed to tap 4*

1 Selector movement : Selector 2 moves to lead number 9
: Selector 1 set at lead number 8
: Pre-selector set between 3 and 4

2 Diverter movement : Existing feed through M_1

a) Bridges M_1 and A_1 → Load current passes through M_1 and A_1 from lead 8

b) Released from M_1 → Load current passes through A_1 only

c) Bridges A_1 and A_2 → Load current divides on lead 8 and 9. Bridging of lead 8 and 9 is taken care of by R_1 and R_2 acting as load for circulating current

d) Released from A_1 → Load current passes through A_2 only (from lead 9)

e) Bridges M_2 and A_2 → Load current passes through M_2 and A_2 (from lead 9)

f) Released from A_2 and settles at M_2 → Load current from lead 9 passes through M_2 i.e., tap set at 4.

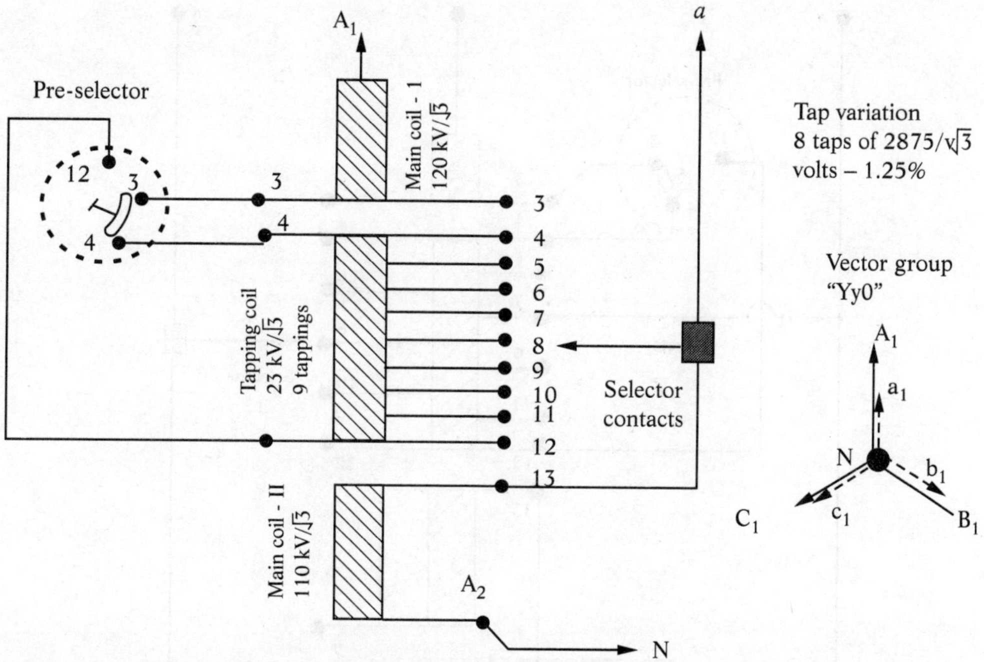

Fig. 1.60 *Tap operation of 100 MVA auto transformer 230 / 110 kV*

1.5.6 Mechanism for On-load Tap Changers

It is a fundamental requirement of all tap changers that the tap selector and diverter switch combination operate in the correct sequences described above. The desired alternating movements are achieved by employing either

(i) Geneva wheel, one on each spindle of a tapping selector with a common operating wheel in the centre that operates each selector shaft in turn, or

(ii) with an elliptical gear train. The central spindle is driven at constant speed and the outer spindles operate alternatively completing about 330° operation in turn for each 180° movement of the centre spindle. Thus, alternate motion of the two selectors is obtained.

An extension to this gear train is also available for the movement of a change-over selector. The tap changer can be maintained under a head of oil provided by the conservator and thereby brought within the zone of a separate gas relay protection. The diverter switches are normally housed below the selectors in a free oil level compartment with inspection covers for maintenance purposes.

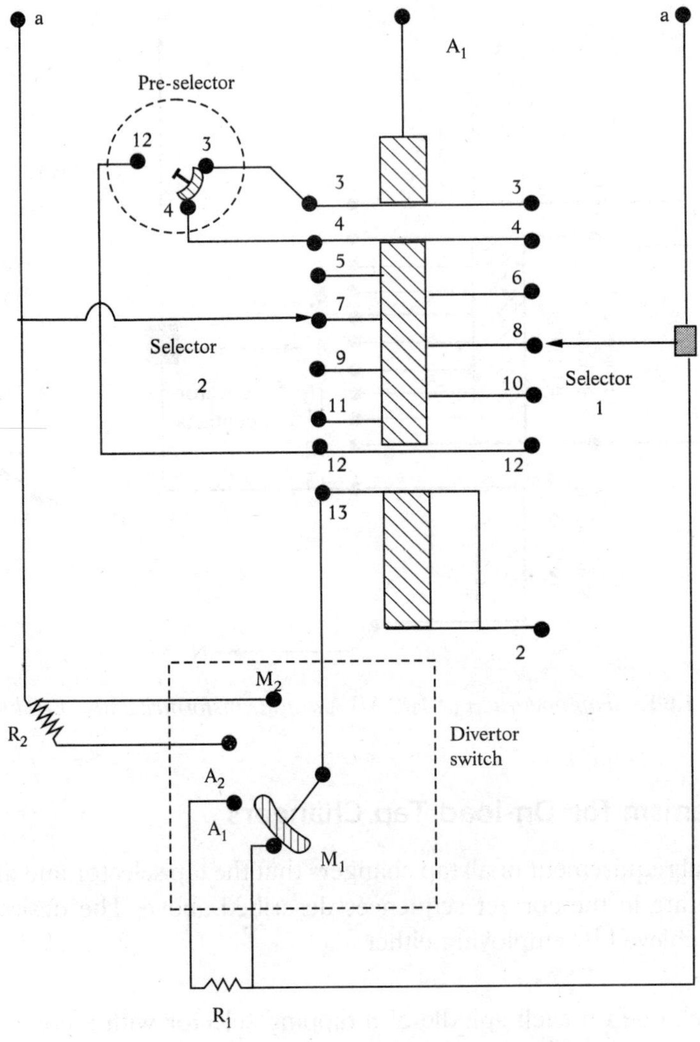

Fig. 1.61 Tap operation 230/110 kV auto transformer

Another arrangement is to mount the tap selectors in compartments at the phase centres along with transformer tank with diverter switches separately housed at the end. This arrangement will have the advantage of short tapping leads. A typical double compartment tap changer is shown in Fig. 1.62.

Control of on-load tap change gear

(a) The control motor is run up in one direction for a 'raise' tap change and in the reverse direction for a 'lower' tap change.
The mechanical gearing interlocks the selector and diverter switches such that once

pressed for tap change, it maintains the sequence of operations to be performed.

(b) Either a break is employed to bring the motor to rest, or clutching and declutching is carried out electrically or mechanically and arranged so that only one tap change at a time is performed for each initiation. Arrangement is also be made to prevent overshoot.

(c) The mechanical arrangement is such that when a pulse is given for a change and it starts the operation, the operation will be performed definitely with the stored energy even if the supply fails. This is called 'go' or 'no go' mechanism.

(d) Mechanical hand operation is always made available for emergency.

(e) Automatic voltage control at substations can be arranged so that a predetermined constant or compensated bus bar voltage can be maintained.

A potential transformer, energised from the controlled voltage side, energises a voltage sensitive relay with contacts to start a tap changer in the required direction.

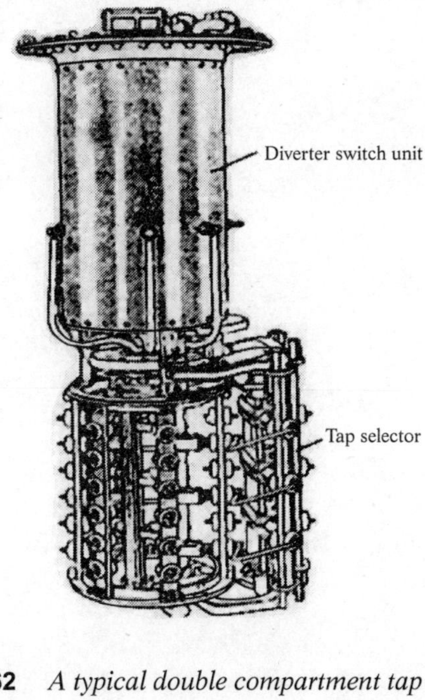

Fig. 1.62 *A typical double compartment tap changer*

A time delay element will also be introduced to prevent unnecessary operation or hunting of tap changer during transient voltage changes.

(f) Line drop compensator may also be added to arrange for an increase in the voltage to be maintained with increase in load. The device comprises a combination of a tapped resistor and tapped reactor, fed from the secondary of a current transformer whose primary winding carries the load current and is arranged to subtract a voltage drop proportional to the load on the main transformer from the applied voltage to the relay.

(g) In parallel operation of the transformer, it is necessary to keep all the transformers in the same tapping position, as otherwise a circulating current will be set up. To achieve this a master–trailer system is followed.

In this, the first tap changer of a group is initiated from an automatic relay and on completing its tap-change, passes on an impulse to the second tap changer which in turn passes an impulse to the next unit as required. Lock-out will be provided so that the first unit cannot operate again until all the groups have come into the line. This system involves auxiliary step-by-step master and follower switches and somewhat complicated interconnections.

The second method is simultaneous operation. Here, all the tap changers are arranged to start their operation at the same time. This is a simple arrangement. It is necessary to provide lockout arrangement to take care of any individual failure (out of step).

2

Transportation, Erection and Commissioning of Power Transformers

This section covers the procedure and precautions to be observed during the transportation, erection and commissioning of power transformers. Detailed descriptions of tests to be carried out before commissioning of power transformers are also described.

2.1 Transportation, Erection and Commissioning

2.1.1 Introduction

The transportation, erection and commissioning of the heavy power transformers for installation is a very important for ensuring their satisfactory placement and service worthiness at the required location to feed loads. This section deals with two categories of transformers: (i) low weight transformers to be distributed and (ii) heavy weight units to be placed in substations.

2.1.2 Low-weight Transformers

The distribution transformers, generally ranging from 50 kVA to 250 kVA, are of lesser weight and are usually mounted on structures or on concrete plinths. A mounting structures is double pole (DP) with a pair of *mild steel seating channels* of size 100 mm × 50 mm, placed at sufficient height from the ground, on which the transformer is housed. In case of plinth mounting, a concrete plinth is constructed from the ground level in between the double poles of the structure, up to a height of 1.5 m (5 feet), and the transformer is placed on it.

For such a placement, the transformer will have its bottom fitted with two small channels of size 75 mm × 27 mm welded to the bottom of the main tank, parallel to the shorter side of the tank so that lesser weight transformers can be placed on the seating channel or on the plinth.

The transportation of the transformer to the site can be done by a lorry. To erect, the procedure is a to lift the transformer from the laden lorry through a chain block hung over the top of the structure, or to move it to the plinth from the lorry in case of plinth mounting.

Distribution transformers are not put to rigorous test during commissioning. Only IR values and oil samples are tested for BDV test before commissioning.

2.1.3 Handling of Heavy-weight Transformers

The power transformers to be installed in substations have large capacities of 10 MVA to 25 MVA in the case of radial 110/11 kV or 22 kV substation, about 50 MVA to 100 MVA in the case of 220/110 kV grid stations, and about 350 MVA in the case of a 400/220 kV substation. The total weights of such transformers are given below to give an idea of the weights involved in transportation and erection.

10 MVA, 110/11 kV transformer	:	90 tonnes
100 MVA, 220/110 kV transformer	:	150 tonnes

The transportation and erection of such heavy units are to be handled with care. The methods of handling are described in detail in the ensuing paragraphs.

2.1.4 Transporting the Transformer

The job of transporting the transformer may rest with either the purchaser or the supplier. In case the supplier delivers it at the location of its utilisation, its handling and safety aspects during transportation and erection are the supplier's concern. Else, the responsibility is of the engineers at the utility (purchaser). The following considerations regarding the safety aspects should be taken care of.

The transportation may involve the following:

(i) Transportation of the unit from one place of utilisation to another in case of change of capacity

(ii) Transportation of the unit from a railway wagon in case the consignment is sent by the manufacturer to the nearest railway station where sidings are available

(iii) Transportation of the unit from the dock to the destination in case the consignment is shipped by the manufacturer to a nearby port

The consignments may be of the following modes (conditions) depending on their weights and dimensions.

(a) Completely assembled, oil-filled and ready for service

(b) Assembled in the tank, filled with enough oil to cover the core and coils, with the remaining oil filled in drums and the other accessories intact with the main tank

(c) Core and windings oil-immersed, but other accessories removed and the remaining oil in drums

(d) Core and windings kept sealed in tank with dry air or inert gas under pressure. Oil and other accessories are sent separately.

Thus, before accepting the consignment through railway receipt or ship document, all the individual components packed in wooden crates should be checked with along with their weights and sizes.

2.1.5 Dismantling of Accessories

In case the transformer is being transported from one place of utilisation to another, the accessories should be dismantled before transportation. The main tank comprising the core and the windings, filled with oil immersing the core, is to be transported separately. This will reduce weight during transportation.

The following accessories are dismantled.

1. Cooling pipe / radiators after closing their header valves
2. Conservator, its pipe arrangement and Buchholz relay
3. HV and LV bushings
4. Equviliser pipe
5. OLTC mechanism box and its tandem operation arrangements
6. Temperature indicating systems

After dismantling the accessories, their corresponding holders are to be closed with dummy plates of required sizes through gaskets to avoid entry of suspended impurities and moisture.

2.1.6 Loading Transformer on Trucks

For very heavy transformers, a truck hauled by a separate engine cabin unit is preferable, with turn table for negotiating curves in case of long chassis. An 8-wheel or 12-wheel trucks may be chosen depending upon the actual loading on the wheels. The truck's body, if not very high from the ground, will be convenient for loading and unloading.

The ease of access of the truck or lorry to the siding has to be checked and ascertained. Crane facilities for lifting and placing the unit in the truck should be arranged for by the authorities and utilized in the transshipment if necessary.

2.1.7 Precautions to be Taken in Transportation

It is customary to have a pilot lorry precede the truck carrying the main unit during its movement in order to carry some of the accessories, tools and working personnel to pilot the movement. The route to be taken will have to be assessed beforehand to check whether the movement can be accommodated on it, i.e., the strength of the road, its width, negotiating curves, permitted loading on bridges, etc. If the weight to be transported is very high, highways authorities may need to be consulted and permission obtained to move on the selected route. Besides weight consideration, the height of the consignment should also be taken into account. The general permissible maximum height in road transportation is 4.25 metres (14 feet) from the road surface. Though overhead electricity wires and telephone cables are expected to be maintained at their standard heights, there may be sagging at some places. Hence, *kitties* (triangular frame made of wooden sticks mounted on the top of a bamboo pole) will have to be kept ready for use by the personnel in the tractor / pilot lorry to lift the wires for movement of the truck without interruption.

The wheels of the transformer are removed in transportation. The transformer is placed on the truck over wooden planks placed at appropriate places at the bottom to avoid sliding during transportation. The four corners of the transformer are *lashed* with steel ropes to the chassis using *turn buckle* and *bull dog* grippers (tying down the transformer). During movement, the lashings may get loosened and hence personnel should be available on the tractor for checking the lashings and tightening them when required.

If the tractor trailer has a turn table at the back, personnel should be posted for operating the turn table when necessary. Traffic congestion periods should be avoided. The journey should be steady and slow (a speed of 25 km per hour is preferred). Negotiation of *curves* should be done carefully in case of short bends. Sufficient *checker plates* (M.S plates of large thicknesses) should be carried in the lorry for use in case of the wheels getting stuck in loose soil.

2.1.8 Unloading the Transformer on the Plinth

The length and breadth of the transformer will have to be ascertained before constructing the concrete plinth for placing the transformer at the substation. The total weight of the transformer will be considered for designing the plinth. Rails will have to be embedded on the plinth based on the spacing provided for the wheels in the transformer.

It is customary to place the transformer on wheels on the rails. A stop block angle is welded on the rails near the wheel to prevent inadvertent movement in case of heavy short circuit forces felt by the transformer. For unloading the transformer, use of heavy cranes may not be possible because of bus bars and other live equipments stationed nearby. Hence the transformer will have to be unloaded on a platform built up of *wooden sleepers*.

The wooden sleepers should be made of seasoned teak wood to withstand the heavy load of the transformer. The dimension of a sleeper is 3 metre length (10 feet) with

200 mm × 100 mm (8″ × 4″) cross-section as shown in Fig. 2.1. The 100 mm section is used for raising the height of the platform and 200 mm section as the base. The sleepers are stacked to the required height by placing 2 or 3 columns for the width of the platform and is raised to the required height by placing the longitudinal side of the sleepers in rows and columns decided by the height of the truck (Fig. 2.2). Two rails are placed on the top sleeper rows at a height, levelled to move the transformer through the rails from the truck onto the platform, as shown in Fig. 2.2. The bottom channels of the transformer will move perpendicularly on the rails.

The sleeper platform has to be prepared near the tractor to facilitate the movement of the transformer from the tractor to the platform. The rails placed will have to be inserted fully onto the tractor, under the transformer covering its whole length. The transformer is loaded on these rails using two screw jacks placed at the side ends of the transformer.

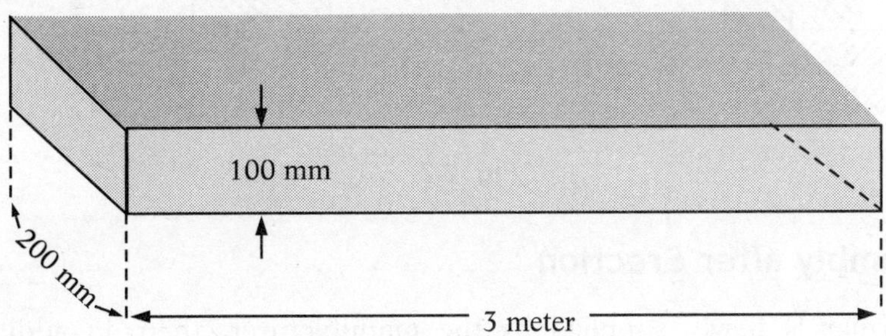

Fig. 2.1

After ascertaining that the transformer is well placed on the rails, a *pull lift* or *tirfor* is attached to the transformer through a D shackle. The pull lift or tirfor will be tied to a cross rail placed on the edge of the top stage sleepers of the platform in order to exercise pulling power.

Now, by operating the pull lift or tirfor, the transformer slowly moves on the rails and is brought to the sleeper stage platform. Then, using screw jacks the rails are removed and the transformer is placed on the sleeper stage. The sleepers will then be removed stage by stage using screw jacks. During removal of the stage, raising the transformer more than the required level will topple the transformer. Hence, utmost care is to be exercised in removing the sleeper stages one by one till finally all the stages are removed. The transformer is then placed again on the rails on the ground for moving to the plinth. Using pulling equipment, the transformer is brought to the plinth and placed properly on the embedded rails of the plinth on its wheels and the stop blocks are then welded.

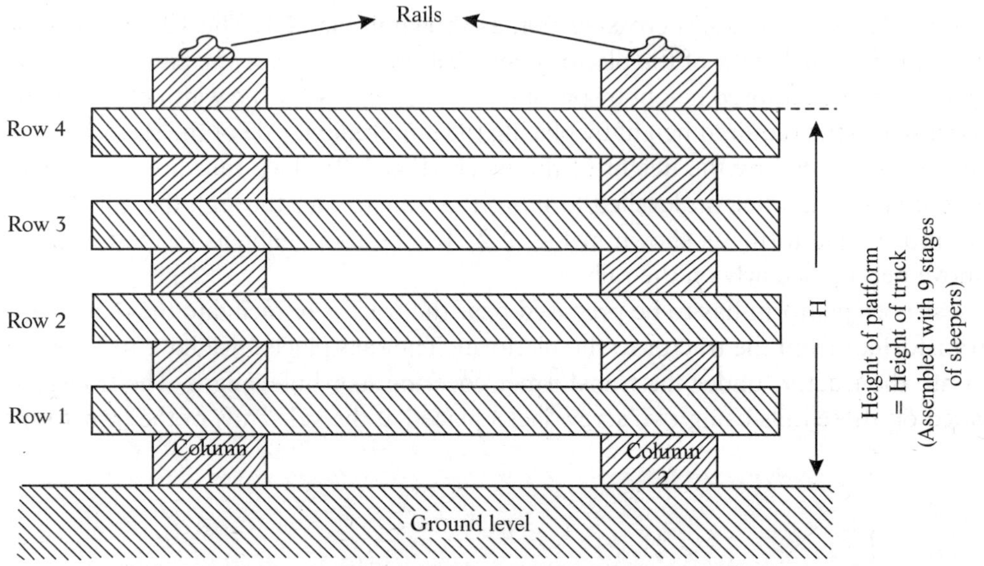

Fig. 2.2

2.1.9 Assembly after Erection

If the transformer is newly supplied by the manufacturer, then, in addition to the dismantling of accessories, all the oil is also despatched separately keeping the core coil unit in the main tank filled with inert gas (nitrogen), properly sealed to prevent entry of moisture.

Hence, the first task in assembling is to circulate and recondition the oil through the filters. Hot oil (70° C) is filled in the main tank, duly replacing the inert gas by letting it go out from the top when oil is pumped in from the bottom through the filter. The hot oil is allowed to fill just below the top cover, completely immersing the core coil unit. A vacuum pump may be used to flush out nitrogen.

The dismantled accessories should be cleaned, dummy plates removed and accessories placed in proper position, replacing gaskets if necessary. Such an assembly may be carried out with a single pole derrick, or with the help of a crane-fitted lorry with due safety measures in place if live parts are located nearby.

The final filling of hot oil can then be carried out till the required level of oil on the conservator is reached. If the conservator is provided with bladder / bellows, the same may be inflated with air and then filled with oil. It is preferable to have the conservator side of the transformer kept lifted up during the final oil-filling process so that air bubbles are released and evacuated without leaving air traps inside. If air remains trapped inside, then after energising the transformer it will get slowly released and will accumulate in the Buchholz relay, causing frequent acting up of the *top float*.

The installation of bushings over the turret should be carefully handled. Since weight is concentrated in the bottom portion, the lifting procedure should be carefully managed. The porcelain portion of the bushing should not be subjected to any pressure due to weight. The procedure is shown in the picture in Fig. 2.3.

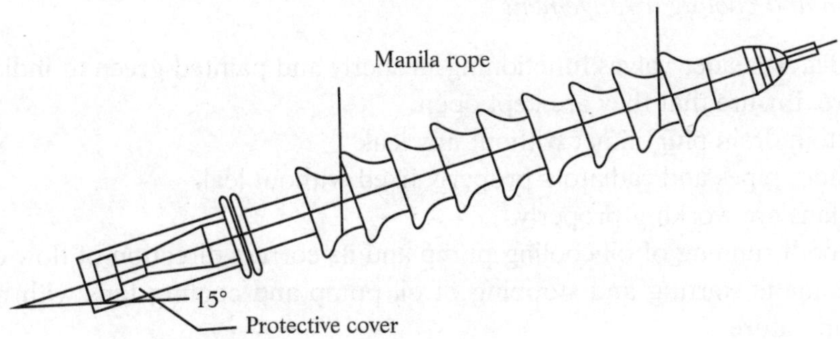

Fig. 2.3

In case of capacitor bushing, tan δ test should be checked before erection of the bushing and also before the commissioning of the transformer.

2.1.10 Hot Oil Circulation

After complete assembly of transformer, hot oil circulation as described in Section 3.3.4 has to be carried out and standard IR values obtained.

2.1.11 Pre-commissioning Test

After erection of the transformer at the appropriate location, complete pre-commissioning test described in Section 2.2.0 should be carried out. This should be kept as the basic history of the transformer, which will be used to judge the working condition of the same later. Also, complete tests on oil including DGA test should be conducted as will be described in Section 3.2.5, which is also kept as a record for future guidance.

A history card for each transformer should be maintained with details regarding its location, repairs and other particulars.

2.1.12 Checklist for Commissioning of Power Transformers

A complete check has to be carried out before commissioning a power transformer. To avoid any omission, the following check list is given for carrying out systematic checks.

1. *Main tank and valves*

   (a) Complete painting of the transformer

(b) Availability of name plate
(c) Drain valve, top and bottom filter valve, etc. N/C valves properly fitted and painted red
(d) Transformer placed on wheels and stop block welded

2. *Radiation and cooling arrangement*

(a) Radiator header valves functioning properly and painted green to indicate N/O valve. Ensure that they are kept open.
(b) Bottom drain plug intact without any leak
(c) Header pipes and radiators properly fixed without leak
(d) All fans are working properly.
(e) Smooth running of oil cooling pump and its correct direction of flow ensured
(f) Automatic starting and stopping of oil pump and cooling fans with respect to temperature
(g) Proper equipment earthing of motors and fans

3. *HV bushings and LV bushings*

(a) Bushings properly fixed without any leakage
(b) Check for tightness of connections – flexible free end clamps used to avoid strain on the bushings
(c) Adequacy of oil level in case of non communicating or condenser bushing
(d) Check rod gaps and ensure correct prescribed gap distances.
(e) Releasing of air in bushings if provided
(f) Satisfactory $\tan\delta$ test for condenser bushings before and after erection
(g) Check up of the cover of the secondary terminals of CT's in bushings
(h) Proper clearance of the bushings and rod gaps with body of the transformer in accordance with the standard

4. *Tap changer (on load)*

(a) Oil level for OLTC and testing of OLTC oil sample
(b) Checking up tap of voltages for all taps
(c) Checking up of operation for
    (i) Tap movements and indications
    (ii) Proper voltage ratio on tap percentage of voltage
(d) Operation through electrical remotes, local and manual operation

5. *Conservator, breather and oil level indicators*

(a) Specified level of oil
(b) Functioning of oil level indicators
(c) Free passage of air
(d) Oil seal provision at the bottom of breather

(e) Dried silica gel provided intact to receive air through gel only
(f) Operation of the conservator valve (N/O painted green) and ensuring open position

6. *Vent pipe / PRV*
   (a) Condition of bottom and top diaphragm checked
   (b) Gauge glass of vent pipe clean and visible
   (c) Function of PRV checked and also its control circuit
7. *Thermometer*
   (a) Calibration of thermometer
   (b) Setting up of the thermometer for oil and winding temperature
   (c) Checking the trip and alarm contacts of the thermometer and its control circuit
8. *Buchholz relay*
   (a) Check the position of the relay
   (b) Check the operation of the top and bottom float by pumping air
   (c) Check control circuits of the above operation
9. *Testing of oil samples (all tests) and checking that the same are within prescribed limits*
   (a) Main tank bottom sample
   (b) Main tank top sample
   (c) OLTC sample
   (d) Conservator sample
10. *Insulation of windings*
   (a) Observe the IR value obtained during oil circulation and its cooling cycle reading
   (b) Check IR value through power operated megger for HT to LT, LT to earth, and HT to LT; perform continuity test
   (c) Check values of all pre-commissioning tests
11. *Earthing*
   (a) Check the neutral bushing earthing with double earth and with adequate earthing strip
   (b) Double earthing of main tank, rails, earth link of top cover and main body
   (c) Proper link of the above with earth pipe and earthing system
12. *Surge diverters*
   (a) Surge diverters are provided on HV and LV side of transformer, within permitted electrical distances

(b) Proper earthing of these surge diverters with the main earthing system

13. *Protection scheme*

(a) Pre-commissioning test on CTs and protection schemes
(b) Check functioning of protection scheme and its control circuit, OLTC panels, etc.

2.2 Pre-commissioning Test on Power Transformers

2.2.1 Introduction

The power transformers, after their installation on a plinth in the substation, have to be necessarily tested before commissioning since they would have been subjected to vibrations during transport. Moreover, the correctness of the name plate details also should be checked for their effective operation along with other transformers in the substation. The tests that are usually conducted on power transformers before commissioning are explained here.

2.2.2 List of Pre-commissioning Tests

The following pre-commissioning tests are conducted.

(a) Ratio test
(b) Short circuit test
(c) No load current test
(d) DC resistance test for windings
(e) Phasing out test for paralleling
(f) Vector group test
(g) Insulation resistance test

If the transformer being tested was in service previously, the test values already recorded can be compared with the present test values and its health adjudged based on that. If it is a new one, it may be ascertained whether the test values obtained are in accordance with the standard values and specifications given by the manufacturer. The above tests are explained in detail next. For conducting the tests, a power transformer with 5 taps of +3% to 9% is considered.

2.2.3 Ratio Test

The ratio test is conducted to ensure that the turns ratio of the windings is correct and that the tappings are made correctly. It also checks whether the windings are connected in the required ratio in the case of a double ratio transformer.

The test circuit is shown in Fig.2.4.

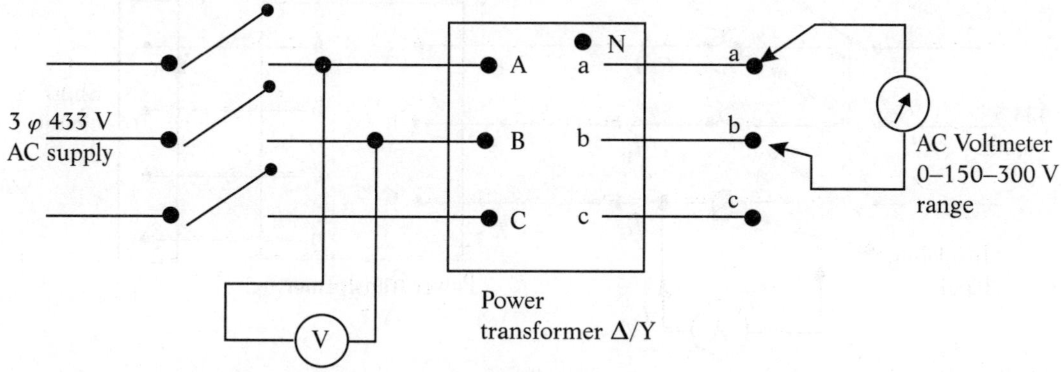

Fig. 2.4

The results obtained by applying 3φ 433 V AC supply on HV side and measuring the voltages on HV side and LV side terminals of the power transformer are tabulated as shown in Table 2.1. This test is conducted in all the tap positions and is compared with the standard value.

When the test voltage is constant, there will be a uniform increase of 3% of voltage measured across the secondary terminals.

Table 2.1

Tap No.	%	HV volts			LV volts						
		AB	BC	CA	ab	bc	ca	an	bn	cn	
1	+3%										
2	Normal										
3	–3%										
4	–6%										
5	–9%										

2.2.4 Short Circuit Test

Three-phase 433 V AC supply is applied across the HV terminals and LV terminals are short circuited with a thick bar of low resistance conductor. When the LV terminals are short circuited, the current limiting impedance of the circuit is only the transformer winding impedance. Based on the percentage impedance value prescribed on the transformer, the short circuit current to flow can be calculated and compared with the test results obtained. The test circuit is given Fig. 2.5.

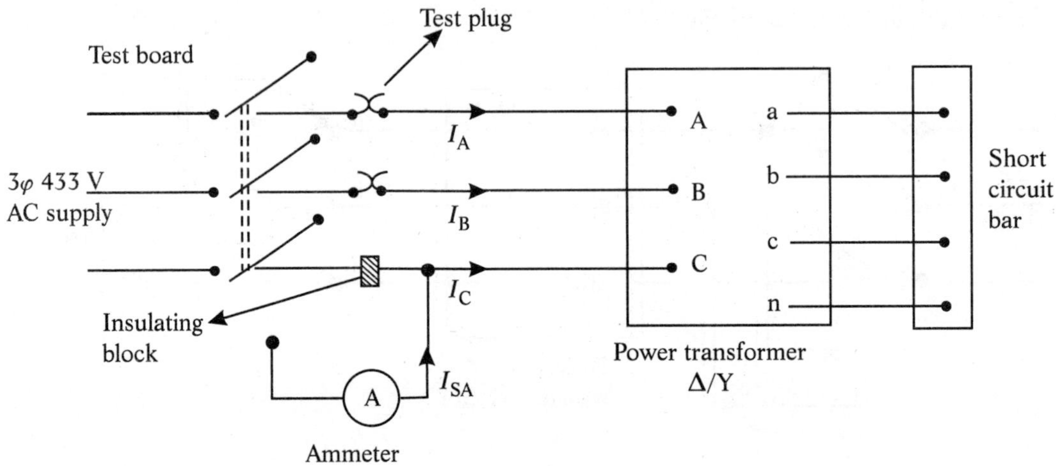

Fig. 2.5

The short circuit current to be passed in the circuit is calculated as below.

$$I_{sh} = I_{rated} \times \frac{V_t}{V_{rated}} \times \frac{100}{Z\%}$$

where I_{sh} = short circuit current obtained by test
I_{rated} = rated current of the transformer
V_{rated} = rated voltage of the transformer
V_t = test voltage applied
$Z\%$ = percentage impedance volts of the transformer.

The results obtained are tabulated as shown in Table 2.2.

Table 2.2

Tap no.	% tap	HV volts			HV currents			
		AB	BC	CA	I_A	I_B	I_C	
1	+3%							
2	Normal							
3	−3%							
4	−6%							
5	−9%							

By this test the percentage impedance prescribed can be tested and hence the health of the windings. Inter turn shorts, improper connections, improper soldering or bracing of cross coil connections in Δ or Y and in tap leads can be ascertained by unequal currents obtained by this test.

By inserting a wattmeter in a phase and taking its reading as P_S with full rated current flowing in the circuit, the full load losses of the transformer can be obtained as follows:

Full load loss $= 3 \times P_S$ watts.

2.2.5 Open-circuit Test

This is done to ascertain the magnetising current of the transformer. The results can be compared with subsequent tests at a later date. Abnormal increase of magnetising current reveals the deteriorating condition of the magnetic circuit (core, core bolts, yoke channels, etc.) The test circuit is given in Fig. 2.6. This test is to be conducted when all the windings of the transformer are included. Hence, in the case under study, +3% tap position includes all windings. The results are recorded as shown in Table 2.3.

Table 2.3

Tap no.	%	HV volts			HV currents			
		AB	BC	CA	I_A	I_B	I_C	
1	+3%							

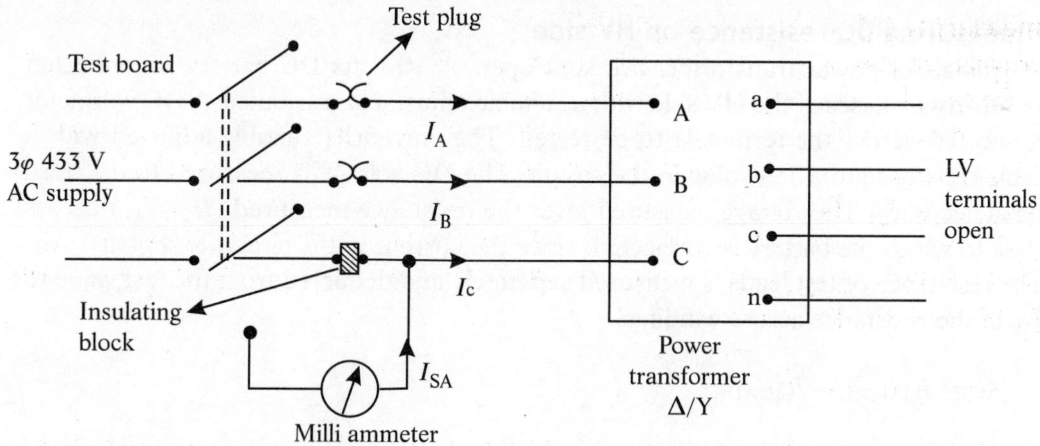

Fig. 2.6

2.2.6 DC Resistance Test

The DC resistances of both HV and LV windings are measured simply by the voltmeter–ammeter method. By this measurement, in addition to the winding resistance, the contact resistance of tap changing contacts, the resistance of interconnecting joints in windings, the lead connection's resistances with the bushing terminals inside the tank are also included. Hence, the health of all these items are revealed by this test.

If an *offload tap changer* is employed, this test is to be conducted whenever the taps are changed to ensure that the contacts of taps are in position and with correct contact pressure. The test circuit is given in Fig. 2.7 for the primary side. For the secondary side, the same test circuit is connected to the secondary side, phase to phase and phase to neutral, with the HV side open.

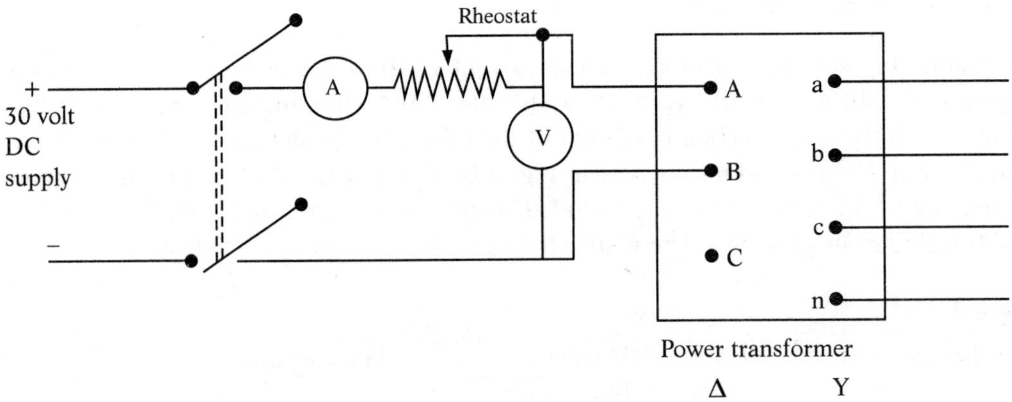

Fig. 2.7

For measuring DC resistance on HV side

LV terminals of power transformer are kept open. A 30 volts DC battery is connected across the two phases of the HV side with a suitable adjustable resistance. A DC voltmeter is connected across the terminals to be tested. The current is suitably adjusted with a variable resistance to a 1 A value in the circuit. The DC voltage is measured to the third decimal accuracy. The voltage measured gives the resistance measured (R_{TOTAL}) across the line to which the battery is connected, since the current in the circuit is kept at 1 A.

The resistance of test leads is measured separately and deducted from the test value to arrive at the resistance across windings.

$$R_W = R_{\text{TOTAL}} - R_{\text{LEADS}}$$

Since the primary is connected in delta, the individual HV winding resistance is calculated from the resistance across the winding as follows.

This test is repeated, two terminals at a time, for all the three lines (A–B, B–C, C–A), and is conducted for all the taps on the HV side and tabulated as shown in Table 2.4. Also, the temperature of the oil and winding should be recorded accurately, since they have a bearing on the resistance values.

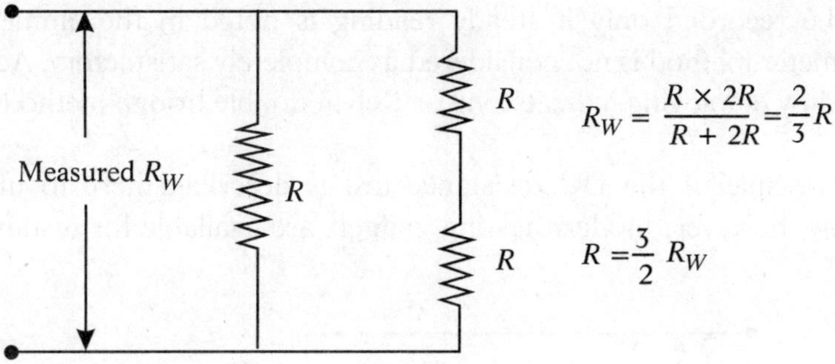

Fig. 2.8

HV side measurement:

(1)	Temperature of oil	=	°C
(2)	Temperature of winding	=	°C
(3)	Resistance of Test Leads	=	Ω

Table 2.4

Tap no.	Percentage	Current Passed	Volts (resistance) measured R_{TOTAL}			Individual winding line resistance		
			A–B	B–C	C–A	A	B	C
1	+3%	1 A						
2	N	1 A						
3	–3%	1 A						
4	–6%	1 A						
5	–9%	1 A						

For measuring DC resistance on LV side

The test circuit shown in Fig. 2.9 is applied on the LV side with HV open. The test is conducted for two lines at a time for all the six combinations, i.e., a–b, b–c, c–a, a–n, b–n, c–n, and the values are recorded.

Since no tapings are available on the LV side, only one test is conducted for all the six combinations.

Here, since the winding is connected in star, the resistance measured across two phases is the combined resistance of two windings in series. The individual winding resistance is the measured value multiplied by 0.5.

This can be compared with the values obtained for a–n, b–n and c–n to check the effectiveness of connection of the winding star point (neutral).

In the above tests, due to high inductive circuit, the DC ammeter will not show any reading immediately when the DC supply is switched on. The DC ammeter will show reading only after sometime, and a gradual increase of readings will be observed. The values are to be recorded only if steady reading is noted in the ammeter. This DC voltmeter–ammeter method is not considered as completely satisfactory. Accurate results can be obtained by employing Wheatstone or Kelvin double bridge method of measuring DC resistance.

The basic principle of the DC resistance test is described here to understand the principle. Today, however, modern testing gadgets are available for instant test without time delay.

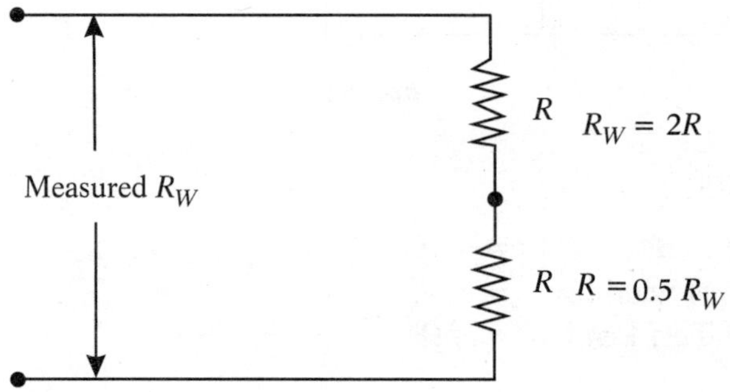

Fig. 2.9

2.2.7 Vector Group Test

Vector Group

In a 3φ transformer, the primary windings can be interconnected in some fashion and the secondary windings connected in some other fashion. This arrangement is called *vector group connection.*

There are four vector group connections and each group has three subgroups. The four main groups are identified by the phase angle displacement of the phase of the secondary winding with its corresponding phase of the primary winding. They are:

Group No. 1:	$0° \rightarrow$	0 hour clock position (12 o'clock)
Group No. 2:	$180° \rightarrow$	6 o'clock position
Group No. 3:	$-30° \rightarrow$	1 o'clock position
Group No. 4:	$+30° \rightarrow$	11 o'clock position

The three sub groups of each main group is identified by the type of interconnections on the primary winding and in the secondary winding while the phase angle difference (o'clock position of vectors) is the same.

For e.g., 4_1DY11 refers to the vector main group 4, subgroup 1, primary in delta, secondary in star, phase angle displacement $+30°$ (11 o'clock position) between identical phases of the primary and the secondary. This connection and the corresponding vectors are shown in Fig. 2.10.

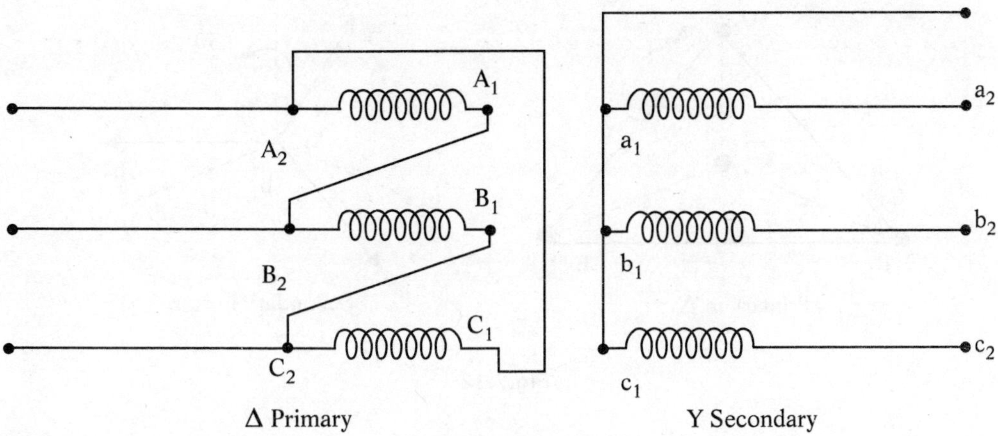

Fig.2.10

Since the primary winding and its corresponding secondary winding is wound on the same core limb, the voltage induced in the secondary winding is identical in phase to the voltage applied to the corresponding primary winding.

The vector of the applied voltage in the primary and that of the induced voltage in the corresponding secondary are parallel as shown in Fig. 2.11when the phase windings are not interconnected.

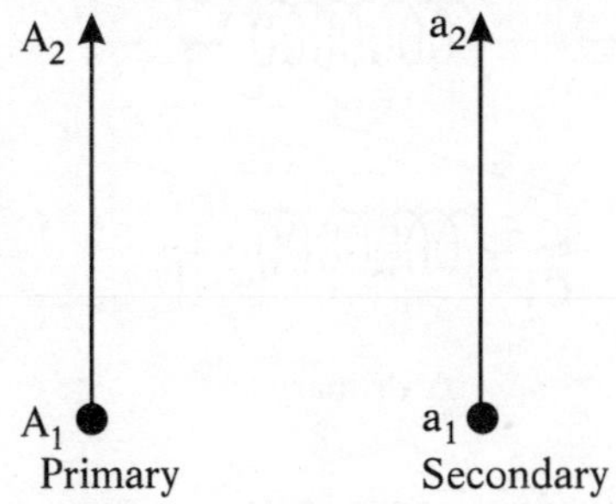

Fig.2.11

When we interconnect the windings to form a vector group, (Group No. 4, here), the phase angle displacement occurs when interconnecting the vectors as shown in Fig. 2.12.

For the Δ vectors of the primary, while drawing the corresponding parallel vectors of the secondary, it forms a tilted star as shown. For the primary delta, a star is shown in dotted lines inside the delta. This star when superimposed on the secondary star vectors (shown in dotted line), a tilting of vectors by +30° is observed. This is the vector group 4_1DY11. The formation of various vector groups has been explained in detail in Section 1.4.

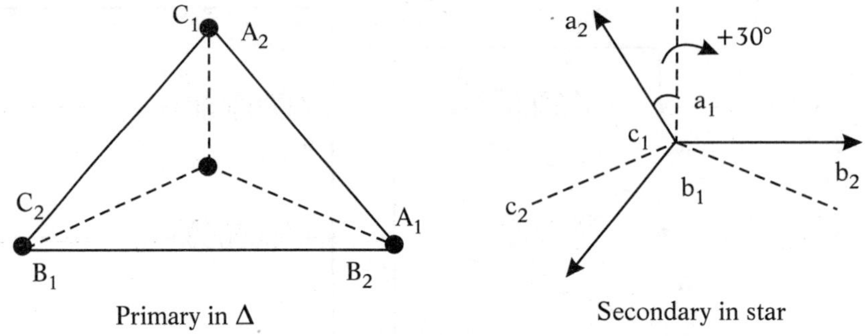

Fig.2.12

Test for vector group

The test is now conducted to identify this vector group 4_1DY11 as shown in Fig. 2.13.

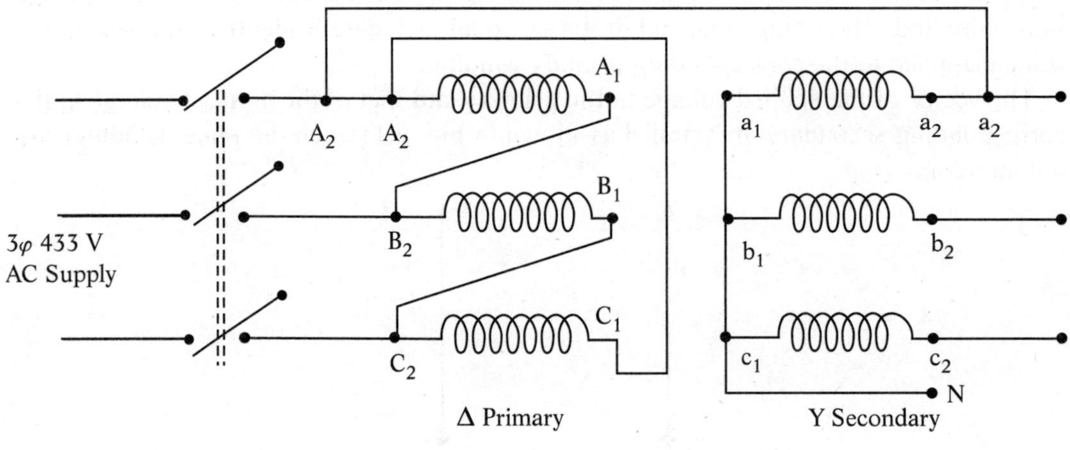

Fig. 2.13

In the test circuit shown, A_2 terminal of the primary is connected to a_2 terminal of the secondary. Thus, one primary terminal is made to coincide with the corresponding secondary. This enables the superimposition of the vector on the corresponding primary vector. LT three-phase supply of 433 volts is applied to the primary and a complete set of voltage reading are taken between possible pairs of the secondary terminals. These readings will determine the dimensions of the vector triangles of the primary and the secondary, superimposed on one another with A_2 and a_2 common as indicated below.

Values of voltages are measured between:

$A_2–B_2$	$B_2–C_2$	$C_2–a_2$	$a_2–b_2$	$b_2–c_2$
$A_2–C_2$	$B_2–a_2$	$C_2–b_2$	$a_2–C_2$	$b_2–N$
$A_2–a_2$	$B_2–b_2$	$C_2–c_2$	$a_2–N$	$c_2–N$
$A_2–b_2$	$B_2–c_2$	$C_2–N$		
$A_2–c_2$	$B_2–N$			
$A_2–N$				

The above is represented in the superimposed vector diagram as shown in Fig. 2.14.

From this we can infer the following.

1. $C_2–b_2$: Greater than supply voltage
2. $C_2–c_2$: Less than the supply voltage
3. $B_2–b_2$ and $C_2–c_2$: Will be equal but less than the supply voltage

The above inference confirms the vector group 4_1DY11.

After superimposing the vector diagram to scale, the measurement taken can also be verified from the superimposed vector diagram plotted.

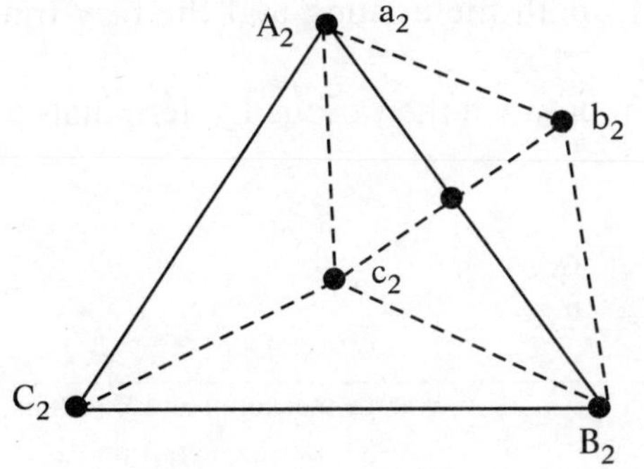

Fig. 2.14

2.2.8 Phasing-out Test

The power transfer in the substation is subjected to parallel operation and hence it is necessary to identify the correct phases with respect to other transformers in service. The correct polarity identification is done by carrying out the phasing-out test.

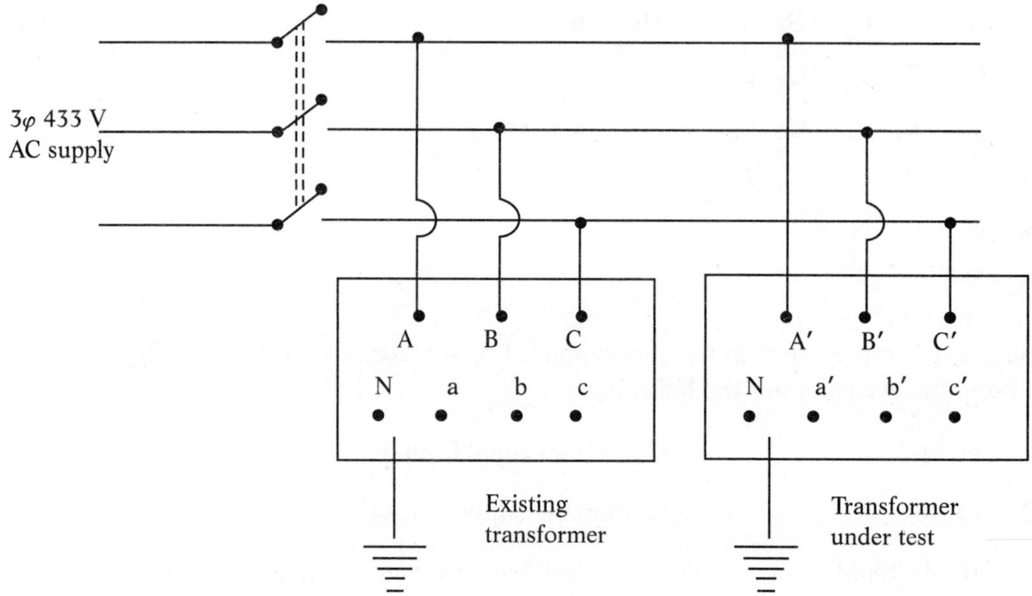

Fig.2.15

Test Circuit

The test is done with taps on both transformers, set in same percentage. Three-phase 433 volts AC supply is given to both the existing and the new transformers simultaneously from the same source (Fig. 2.15).

Measurements are taken between the various LV terminals and recorded as below.

Results

Tap set at transformer 1 = %
Tap set at transformer 2 = %

Test voltage across HV terminal			Voltage measured on LV terminals of transformers 1 and 2													
AB	BC	CA	aa'	bb'	cc'	ab'	ac'	ba'	bc'	ca'	cb'	na'	nb'	nc'	nn'	

If the phasing out is proper, the readings across the terminals a–a, b–b, and c–c will be zero, or a few milli volt difference may occur due to slight differences. Now, we can say the phases are correctly identified with respect to each other. If however, a large difference occurs, the instantaneous values of the potentials of opposite terminals are not the same and this may be due to

(a) incorrect phase sequence
(b) wrong polarity of windings
(c) incorrect connections giving difference in the inherent phase displacement angle between the primary and the secondary windings.

2.2.9 Insulation Resistance Test

The insulation resistance test of the winding is checked by using a 1000 V megger with the neutral earthing removed. The test is carried out for all combinations of phases and earth and values in megaohms are recorded. Since the windings are immersed in oil, the electrical quality of the oil also decides the value of insulation of winding. The temperature of the winding is also a criterion for the *IR* values of the transformer.

The temperature of the transformer oil will be raised to more than 70°C during circulation process. Its *IR* value is obtained during cooling cycle after the circulation for various descending temperatures. The value at 60°C is a critical value, and generally, it should be measured at more than 2 mega ohm/kV. This temperature condition can also be obtained after it is put into service and fully loaded and then switched off.

A curve may be drawn for various descending temperatures from the maximum. The *IR* value at this service temperature condition with 1 minute value to 10 minute value is called polarization index, by which the health of the insulation is also decided.

These tests and interpretations are explained later in Section 3.3.3.

2.2.10 Other Tests Conducted

Besides the electrical tests described above, the following physical tests are also necessary.

(1) Health of the oil decided by carrying out all the prescribed tests
(2) Testing of alarm and trip indications of oil and winding
(3) Testing of the top and the bottom Buchholz relay action and surge relay actions
(4) Testing the action of magnetic oil gauge, PRV, etc.
(5) Testing the action of pressing into service the cooling fans and pump at the set temperatures
(6) Testing for health of the bushings

Apart from the tests described above, the functioning of protective relays, their action at appropriate set values and conditions, and testing of their characteristics are also to be carried out. The health of the transformer is thus decided and the power transformer is put into service.

3

Maintenance of Power Transformers

> In this section, the external accessories of transformer, i.e., oil handling components, bushings and protection components and their constructional features are described. Transformer oil usage, various oil tests which are to be conducted and the reclamation process for oil in case of unsatisfactory oil test results are discussed. The processes of maintenance works and improvement of IR values are focussed upon. Troubleshooting in transformers, the causes necessitating it and the remedial measures to be taken up by the maintenance wing are enumerated.

3.1 External Accessories of a Transformer

3.1.1 Introduction

The use of oil in transformers for both insulation and cooling purposes necessitates the use of some external accessories. These accessories can be classified as (i) components for oil handling systems, (ii) bushings and (iii) protective components. These components are kept attached with the *main tank* inside which the core-coil unit is placed. The accessories are as follows.

1. Oil handling system components
   (a) Main tank with flow valves and top cover
   (b) Cooling pipes or radiators with cooling systems
   (c) Conservator
   (d) Breather
   (e) Oil level indicators
   (f) Temperature measuring systems

2. Bushings for making out connections
3. Protective components
   (a) Buchholz relay
   (b) Explosion vent
   (c) Pressure relief valve (PRV) or equaliser pipe
   (d) Air release plug or cock

The description of all the above accessories with their purposes and constructional features are discussed in this section.

3.1.2 Main Tank with Flow Valves and Top Cover

The core-coil unit has to be placed inside a tank and filled with transformer oil. This main tank with cooling system, oil flow control valves and top cover is shown in Fig. 3.1.

Main tank and top cover

The main tank for housing the core coil unit of the transformer is rectangular in shape, made of boiler plates with welded joints, and is leak proof. At the bottom of the tank, in case of large capacity transformers, wheels are provided to be placed on rails embedded in the concrete plinth, with a stop block welded to it to prevent inadvertent movement while in service. In small capacity distribution transformers, two channels are kept welded parallel to the shorter side of the tank in order to place it in the structure.

The rectangular flange at the top of the tank has holes to allow the fastening of the top cover and gaskets with bolts. The gaskets, used to prevent oil leak, should be of sufficient size to accommodate the weight of the top cover and other accessories placed on it.

Flow valves attached to the tank

(i) **Drain valve**: This is provided at the bottom of the tank to drain the oil after cleaning. Since sludge and moisture will settle inside at the bottom of the tank, this can be used to verify contamination. The oil drained through this valve should not be re-used.

(ii) **Bottom sampling / filter valve:** This is provided at a height of about 100 mm from the bottom of the tank and is used for taking bottom oil sample for test and to connect the oil filter suction hose for filtering.

(iii) **Top sampling / filter valve:** This valve is diagonally opposite the bottom sampling valve and is used for taking top oil sample for test and to connect the oil filter delivery hose for filtering.

(iv) **Header valves for cooling system:** The top and bottom header pipes to which the cooling system is attached, each have a valve with flange to control the flow of oil to the cooling system. These are called the top header valve and the bottom header valve respectively. The oil flow has to be stopped during oil circulation through the filter or in the case of dismantling of the cooling system for transport. There is also be an oil drain plug at the end of the bottom header pipe for draining off the oil from the radiators. The oil flow valves mentioned in item (i) to (iii) are normally closed (N/C), whereas the header valves are kept open (N/O) during the working of the transformer.

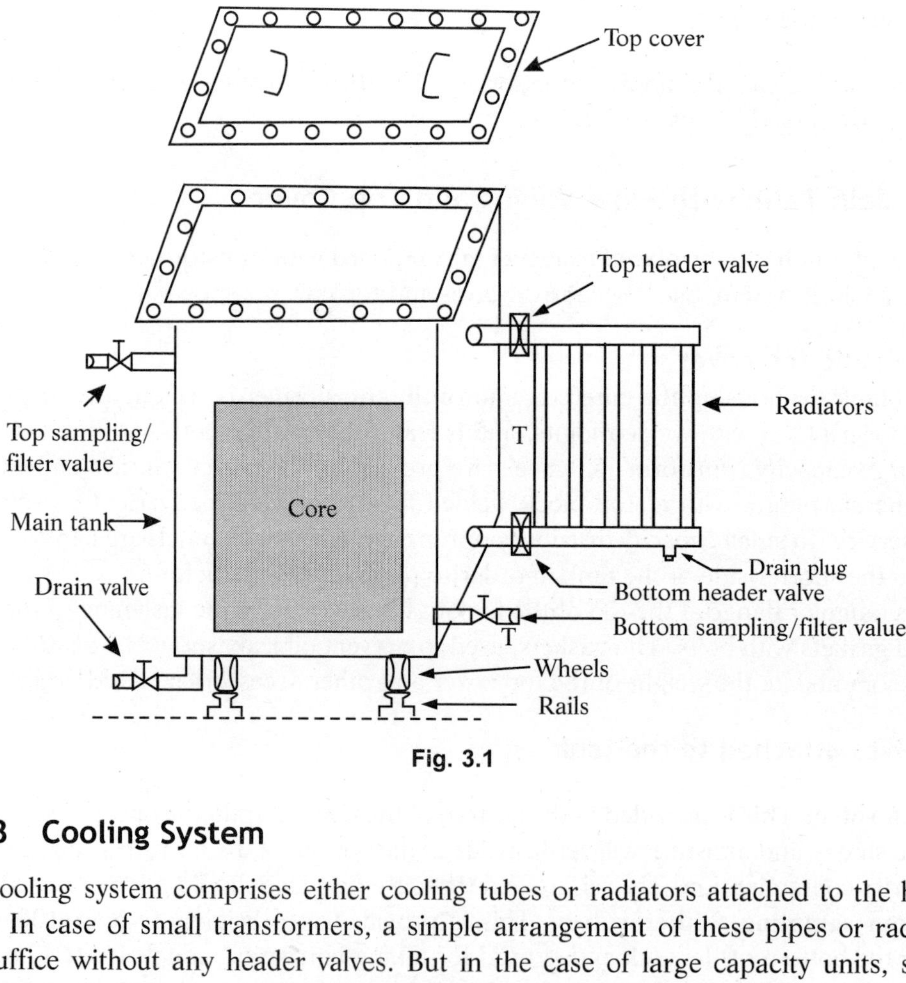

Fig. 3.1

3.1.3 Cooling System

The cooling system comprises either cooling tubes or radiators attached to the header pipes. In case of small transformers, a simple arrangement of these pipes or radiators will suffice without any header valves. But in the case of large capacity units, special arrangements will have to be made for effective cooling as discussed next.

The cooling system, in general, is classified into four types, i.e.,

(i) ONAN - Oil natural cooling with natural air (basic arrangement)
(ii) ONAF - Oil natural with forced air
(iii) OFAF - Oil forced with forced air
(iv) OFWF - Oil forced with water forced

The various arrangements are shown in Fig. 3.2.

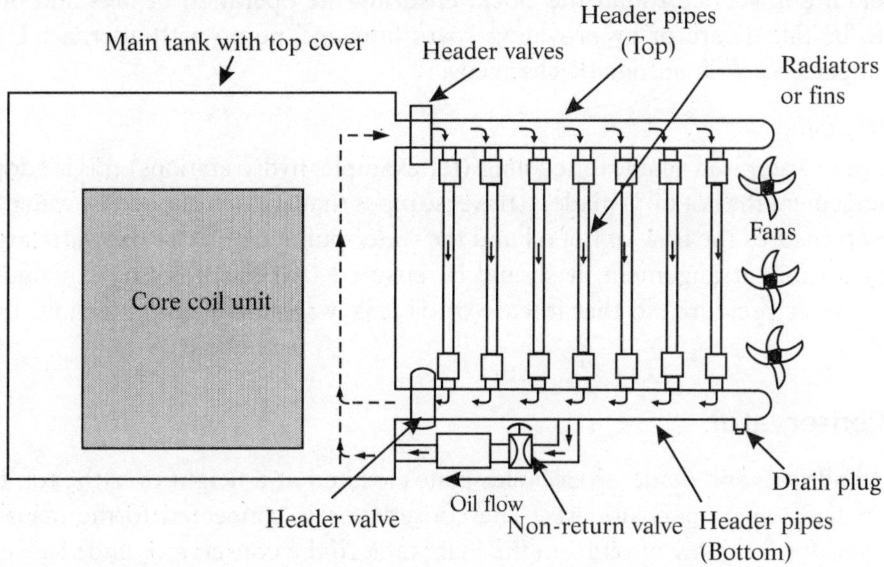

Fig. 3.2

(i) ONAN cooling

This is a basic cooling arrangement available on all types of transformers. The natural tendency of oil is to absorb heat from the core-coil unit and rise to the top of the tank as indicated in the figure. Then it enters the header pipe at the top, flows through radiator fins and gets cooled through natural air. The cool oil enters the transformer main tank through the bottom header pipe. The circulation happens naturally and hence the name.

It can be inferred now that the entry of the bottom header pipe should be near the bottom of the core. There should also be sufficient space from the top of the core to the top cover and the top header pipe should be below the top cover in order to make this natural circulation effective. The hot oil impinges all the sides of the main tank and at the top cover from where the heat is radiated to the atmospheric air.

The process of absorbing heat by the oil from the core is by conduction. The heat is carried away through convection by the oil and is radiated to the atmosphere by radiation.

(ii) ONAF cooling

By providing fans at the radiators, by *exhaust* or *blower* arrangement, the cooling through radiators is accelerated.

(iii) OFAF cooling

In addition to the fans, an oil pump is also provided (submergible motor and pump) with a non return (NR) valve in order to ensure the direction of oil flow from radiator to main tank through bottom header pipe to aid natural circulation. The NR valve opens with a sound when the pump is started with oil flow in the correct direction only. Since power transformers are in service round the clock, ensuring the operation of fans and pumps will have to be taken care of by providing spare fans and pump with alternate LT AC supply arrangements with automatic changeover.

(iv) OFWF cooling

In places where water is available for cooling (for example, hydro stations) this is adopted. In this arrangement the hot oil is made to traverse pipes that are surrounded by water flow. The oil pump ensures the flow of hot oil and the water pump circulates the water around these pipes. In this arrangement, it should be ensured that oil pressure is maintained higher than water pressure, so that in case of oil leak water does not enter into the oil flow system.

3.1.4 Conservator

This is a cylindrical tank made up of boiler plates located at a height over the top cover of the main tank on proper supports. The conservator is connected to the main tank through a pipe for free flow of oil from the main tank to the conservator and vice versa.

Purpose of a conservator

The purpose of providing a conservator is mainly to conserve space above the hot oil exposed to the atmosphere.

The hot oil in contact with oxygen in air will cause oxidation and produce acidific effect. This will corrode the exposed metal parts of the tank. Hence, if expansion and contraction of hot oil is arranged below the top cover itself instead of providing a conservator, the whole top cover space inside is exposed to oxidation and the corroded metal parts will fall down and settle in the windings resulting in failure.

The expansion and contraction of oil is therefore arranged in a separate tank called the conservator, where the exposed metallic surface above the hot oil is considerably less. The corroded metal particles falling down are made to settle at the bottom of the conservator and are prevented from entering the main tank.

The other purposes of a conservator are given below.

(a) It ensures that the core and other accessories immersed in oil because of its position high above the main tank, maintaining a head of oil column.
(b) It enables fixing of the Buchholz relay in the pipe linking the main tank and conservator to ensure protective action.
(c) Expansion and contraction of oil takes place in a separate area, leaving sufficient space for atmospheric contact.

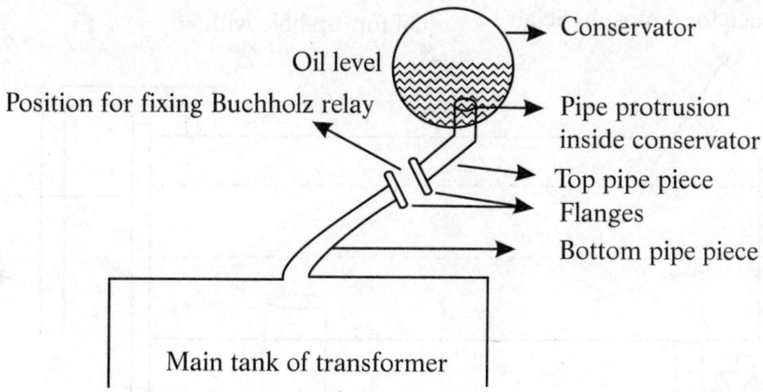

Fig. 3.3

Constructional features

(a) A cylindrical tank of circular cross section and sufficient length is placed in a horizontal position at a height over the top cover, supported by a metal frame and plates.

The volume of the tank depends upon the quantity of oil in main tank, the height of the required expansion column and the space required above the oil column to maintain atmospheric pressure. The volume of oil in the conservator may be 8 to 10% of the main tank oil.

(b) The main tank and the conservator tank are connected through a pipe in which oil will flow freely during its expansion and contraction due to load conditions.

(c) This interlinking pipe is made up of two pieces. The top piece is welded with the conservator bottom and protruded inside the conservator to a length of 75 to 100 mm. The bottom piece is attached to the top cover of the transformer through a flange and a gasket. The other ends of these two pipe pieces also have flanges to enable fixing of the Buchholz relay in between.

(d) The protrusion of the pipe inside the conservator allows the corroded metal particles to settle down at the bottom of the conservator, thereby preventing their entry into the main tank.

(e) One side of this cylindrical tank has a cover fixed to it through gaskets. This is to carry out the cleaning work inside the conservator where the metal particles would have settled at the bottom.

(f) The removable cover also has an arrangement for fixing gauge glass for oil level indication.

(g) On the top of the conservator, a pipe piece with an airtight lid is welded to enable oil top up and for checking the oil level. This top hole is not aligned with the bottom pipe entry but kept offset.

(h) A drain valve is provided at the bottom of the conservator for (i) draining oil and (ii) taking oil samples for testing.

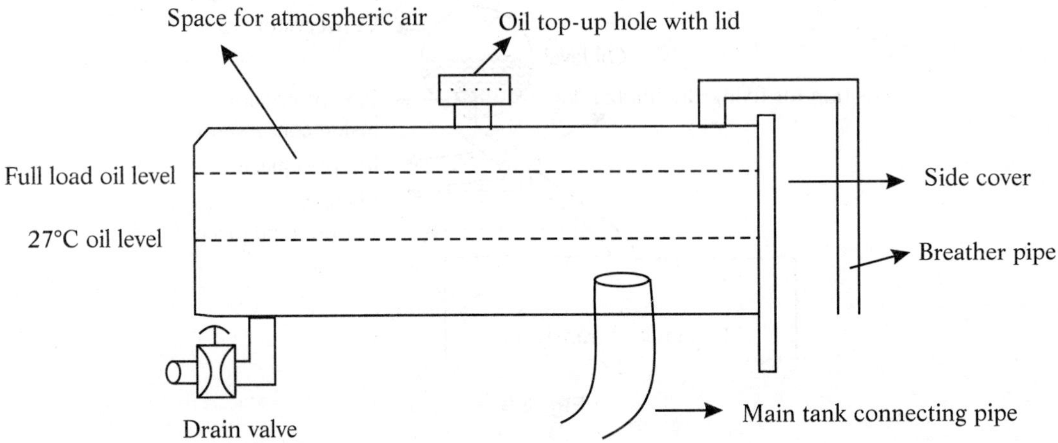

Fig. 3.4

(i) A breather connecting pipe welded at the top of conservator is brought down to the ground level for attaching the breather.

Conservator arrangement for OLTC

A transformer with an on-load tap changer may require a separate conservator for its diverter chamber. For this, either a separate conservator is provided, or else a portion of this main conservator is partitioned and used as separate conservators for the main tank and the OLTC chamber.

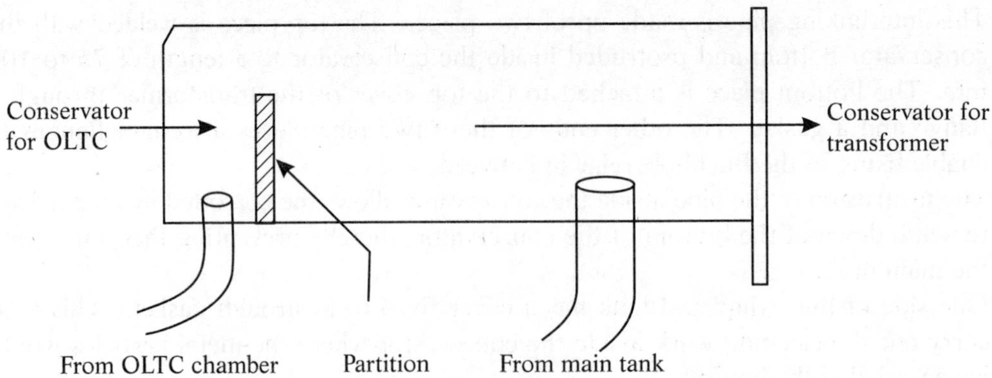

Fig. 3.5

This partition is made at 3/4th the height of conservator such that top atmospheric space is common to both conservators. In case full partition is made, then a separate oil top-up and breather-pipe arrangement are to be provided. A separate conservator for OLTC kept at a height lesser than the main conservator is preferable for detecting oil leakage from the main tank to the OLTC chamber.

Conservator with bladder

Providing space at the top of inside portion of the conservator makes the hot oil come in direct contact with oxygen in the air and results in its getting oxidised to produce an acidic effect. Hence, a bladder or bellow is placed inside the conservator and the breather pipe is kept connected to this bladder/bellow. The volume of air is contained in this bladder/bellow only, and air does not come in direct contact with the hot oil.

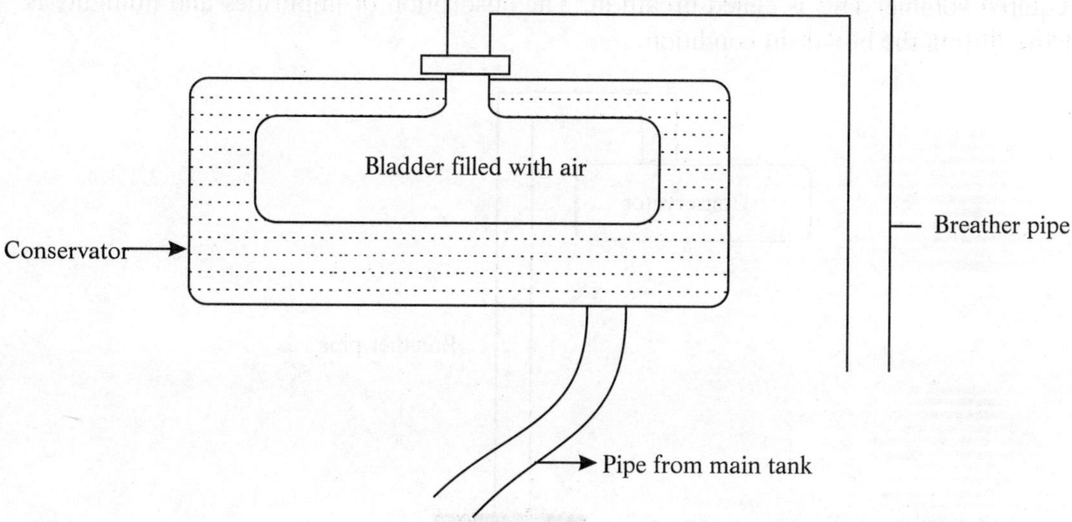

Fig. 3.6

The total volume of conservator is equal to the volume of oil plus the volume of air in the bladder or bellow. If oil expands due to increase of load, it compresses the bladder and sends out some required volume of air from the bladder. If oil contracts due to decrease of load, the required quantity of air is drawn in into the bladder and thus the total volume in the conservator is maintained. This arrangement avoids the contact of oil with atmospheric air, thereby preventing oxidation from taking place. The bladder/bellow should be inflated with air initially, before filling up oil inside the conservator. With this arrangement the topping up of oil should be done only through a filter through any of the flow valves.

3.1.5 Breather

A breather kept attached to the breather pipe is connected to the conservator to keep the atmospheric air pure, without suspended impurities and humidity.

In case of small transformers the breather is attached near the conservator itself. In case of large capacity transformers the breather attachment pipe is extended to the ground level, allowing the breather to be maintained without switching off the transformer.

The atmospheric air inside the conservator has to be free from suspended impurities and moisture. Hence, the breather arrangement comes with an oil seal to absorb suspended impurities and a silica gel container to absorb humidity. Thus, only pure air is allowed inside.

The transformer, when loaded, makes the oil expand in the conservator, *which pushes the air out*. This is called breath out and is observed as bubbles in the oil seal. When the load decreases, the oil level shrinks, and hence air from outside enters to fill up the required volume. This is called breath in. The absorption of impurities and humidity is active during the breath-in condition.

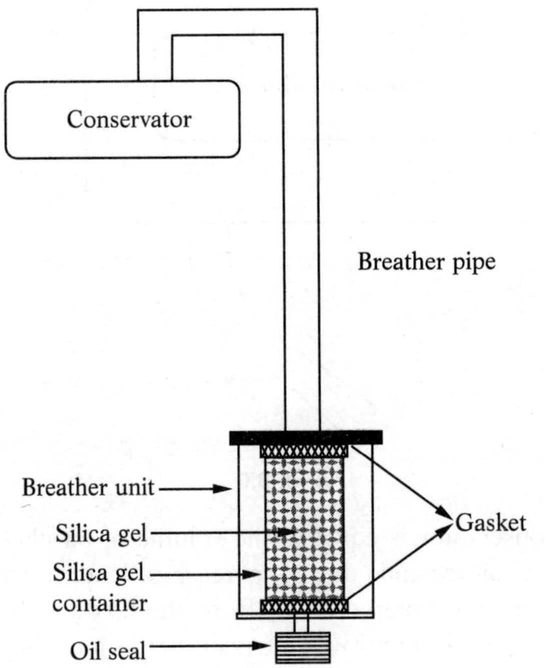

Fig. 3.7

3.1.6 Oil Level Indicators

Generally, three types of oil level indicators are used in transformer applications.

(1) Plain glass type
(2) Glass tube type
(3) Magnetic oil gauge (MOG)

Plain glass type

A plain glass piece is placed on the side of the conservator with a fixing bracket as shown in Fig. 3.8.

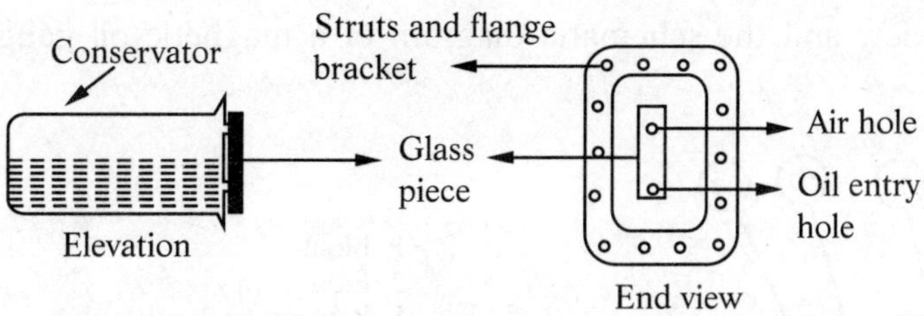

Fig. 3.8

The oil entry hole allows the oil from the conservator into the indicator. The air hole maintains the atmospheric pressure inside the glass. Any blockage in either of the holes will give a wrong indication of the oil level.

Glass tube type

Here, instead of plain glass, a glass tube is fixed for indicating the oil level (Fig. 3.9).

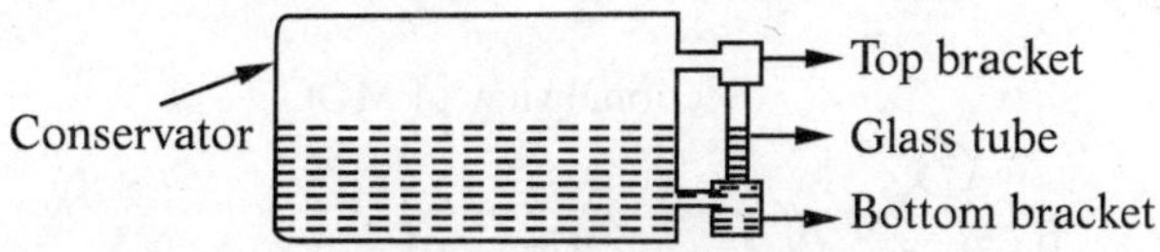

Fig. 3.9

Magnetic oil gauge

In the plain glass type and glass tube type indicators, oil enters the indicator to show the oil level, but in this type, the oil level is measured by a float and the indication given through a driven magnet and a driving magnet. This is explained schematically in Fig. 3.10.

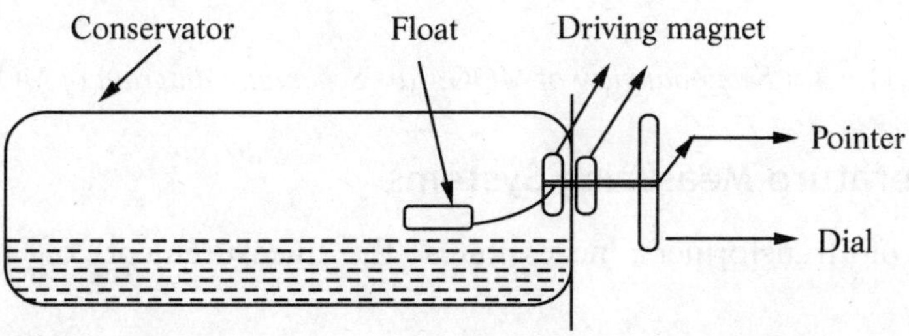

Fig. 3.10

The pointer indication, on going below the required level, initiates a DC circuit that gives out an alarm for low oil condition. Here, any defect in the float, sluggishness in the action of the driving and the driven magnet may cause a wrong indication of the oil level. Hence it is customary to check the oil level physically by opening the top lid of the conservator and inserting a clean dip-stick to make a correct assessment.

The sectional view and the schematic diagram of a magnetic oil gauge (MOG) are given in Fig. 3.11.

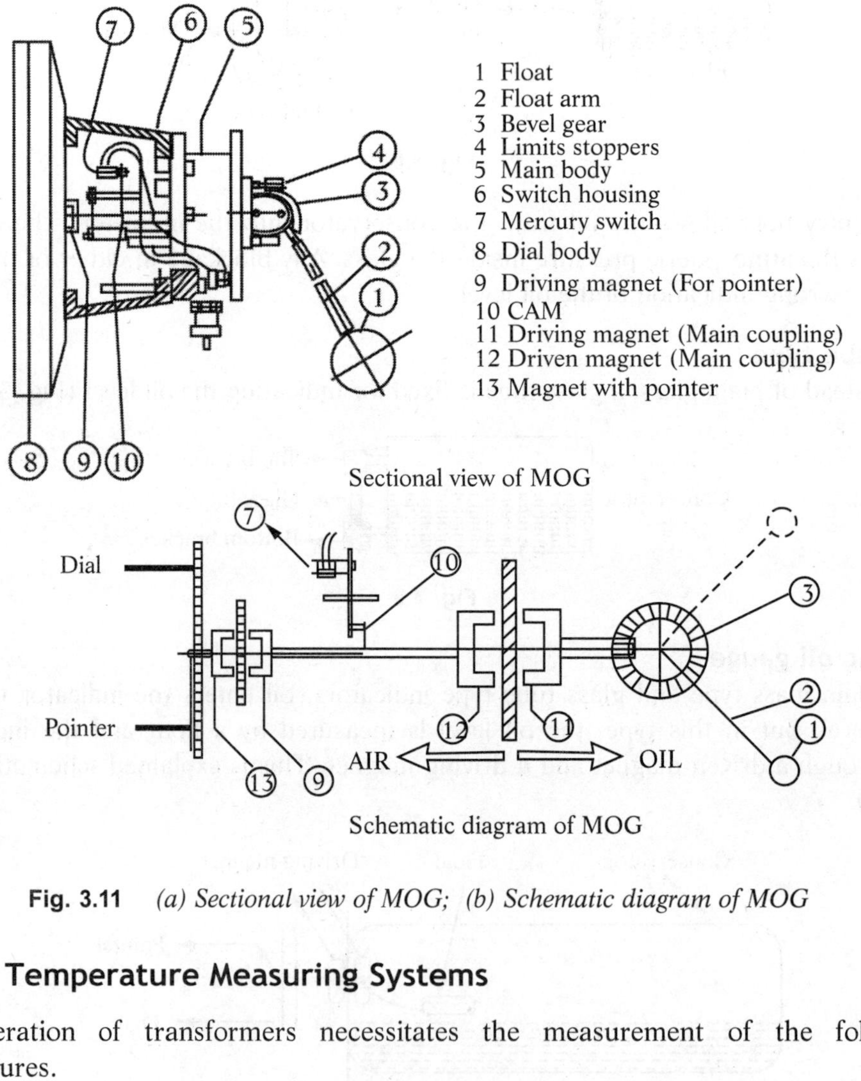

Fig. 3.11 *(a) Sectional view of MOG; (b) Schematic diagram of MOG*

3.1.7 Temperature Measuring Systems

The operation of transformers necessitates the measurement of the following temperatures.

(a) **Winding temperature:** The windings, along with the core, is the place where losses come out as heat and hence it needs to be measured.

(b) **Oil temperature:** The heat generated in the windings and core are carried away by the transformer oil and transmitted to the cooling system. Hence, the oil temperature will have to be measured to ascertain the effectiveness of the cooling system. The oil temperature will always be 5 °C to 10 °C lesser than the winding temperature.

Measuring oil temperature

An oil thermometer pocket is provided on the top cover as shown in Fig. 3.12. This pocket is filled with separate oil and the thermometer stem is placed in it. The heat observed by this stem is communicated to a dial type thermometer through heat insulated capillary tubes. This may also be connected to the special type of thermometers that are currently being used, discussed under the heading *winding temperature measurement* later.

The oil thermometer is set to operate at a specific temperature through a DC circuit to give an alarm indication.

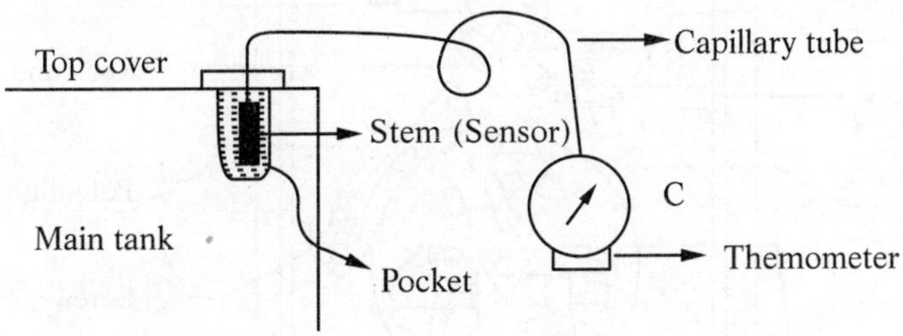

Fig. 3.12

Measuring winding temperature

The temperature in the winding cannot be measured directly by inserting a sensor at the winding itself. Hence, it is measured by the *image* principle by providing a *current transformer* in the winding leads. The secondary current of the CT (max 5 A) proportional to main current is supplied to a heater coil designed to produce a differential heat in conjunction with the oil temperature, and is shown as winding temperature with an appropriate indicator.

Hence, the thermometers for oil and winding are a single unit but with different indicators. The temperature measuring system is shown in Fig. 3.13. The arrangement made for measuring oil and winding temperature is discussed next.

Oil temperature measurement

The sensor stem is placed in the thermometer pocket of the transformer tank. A flexible connecting tube consisting of two capillary tubes is connected to the indicator. One capillary is connected to the operating bellow of the instrument and the other to the compensating bellow.

The operating bellow is filled with a liquid that expands or contracts in volume with varying temperature. The compensating bellow acts upon the operating bellow through a linkage, compensating for variations in the ambient temperature.

The expansion and contraction of the bellows due to temperature variations is transmitted to the indicating pointer and switching disc through a linkage mechanism. Dry electrode mercury switches are mounted on a steel carriage. The make and break setting of temperatures can be adjusted independently for these switches.

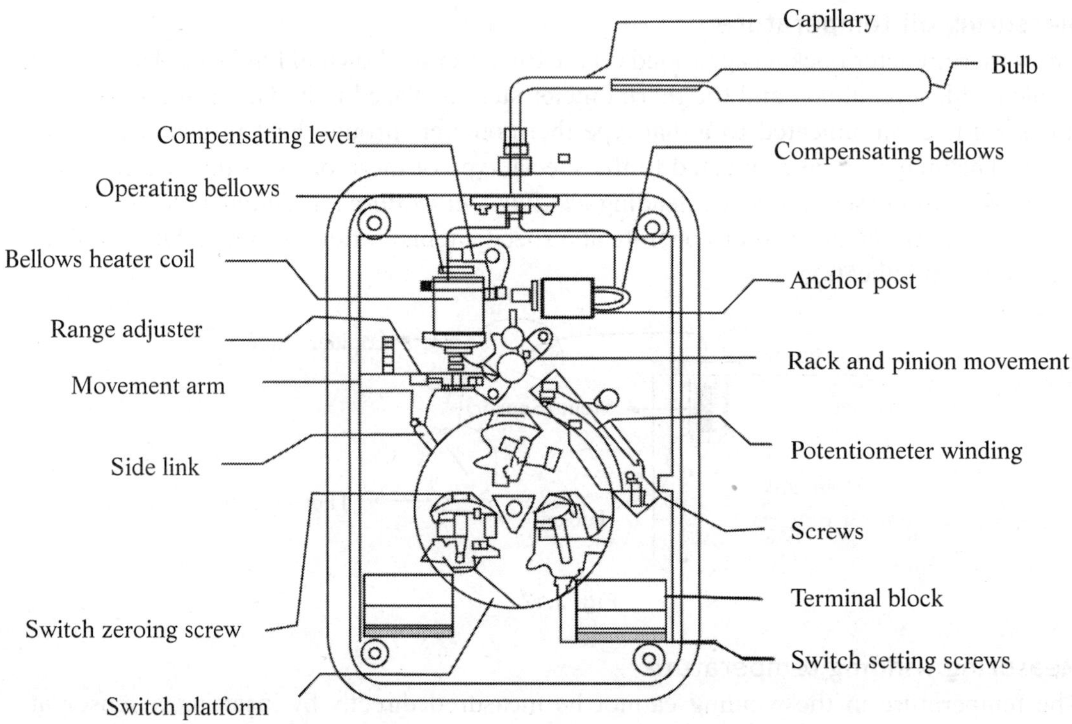

Fig. 3.13 *Sectional view of a temperature indicator*

Winding temperature measurement

The winding temperature indicator is provided with an additional bellow heating element. This heating element is fed by the secondary of the current transformer inserted in the winding leads, with a current proportional to the load. This is arranged in such a manner that the increase in temperature of the heating element is proportional to the temperature increase of the winding over the top oil temperature. The sensor is so placed on the transformer main tank as to enable the measurement of the maximum oil temperature.

The operating bellow thus gets additional movement due to the difference of winding temperature and maximum oil temperature. Hence, it indicates the hotspots in winding temperature.

3.1.8 Bushings

The IN and OUT connections of the supply system has to be made to the windings of the transformer with adequate insulation. This is accomplished by the use of *bushings*.

Plain type bushings

For the LV side of the transformer with voltage levels of 11 kV, 22 kV or 33 kV, a plain type bushing as shown in Fig. 3.14 is appropriate,. This bushing type is a *through rod bushing,* which is externally fixed with collar plate and bolts. The oil is maintained inside the bushing by the head of oil in the conservator.

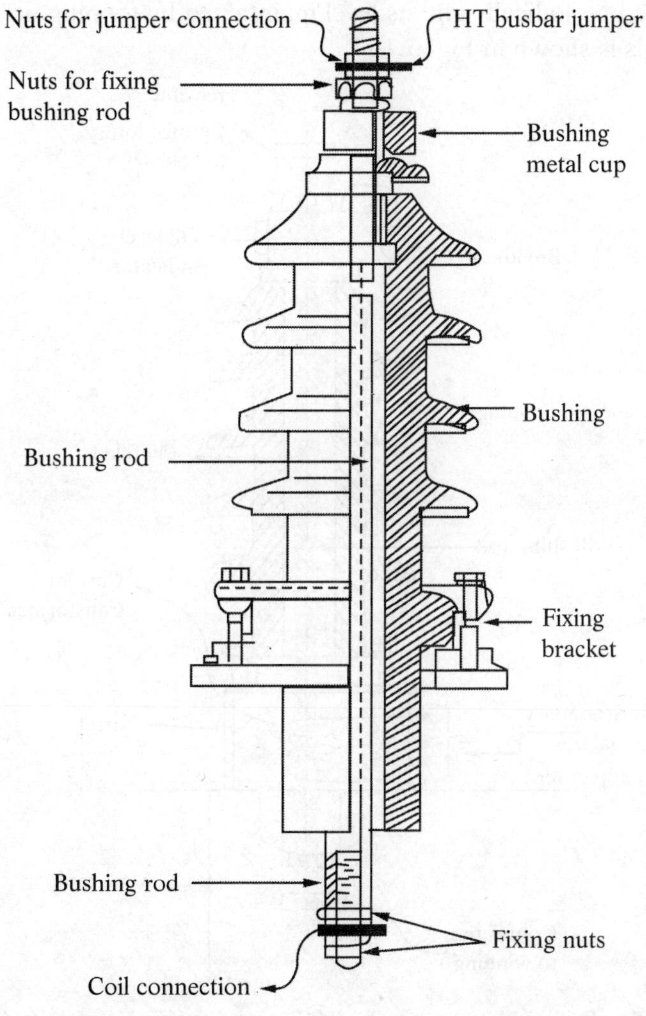

Fig. 3.14 *Fixed HT bushing through the rod half-sectional elevation*

Extra high voltage bushings

Special care should be taken of the bushings housed in the EHV side of the transformer. These bushings are larger and generally, the following three types are employed.

(1) Communicating type oil filled bushing

(2) Non-communicating type oil filled bushing
(3) Condenser type oil impregnated bushing

In the first type, e.g., plain type bushing, the oil in the transformer is communicated to the inside of the bushing. In the second type, i.e., the non-communicating type, oil is maintained in the bushing separately with an oil level indicator and a breather. Therefore, this oil is to be tested periodically and its level maintained by topping up with new tested transformer oil. This is shown in Fig. 3.15.

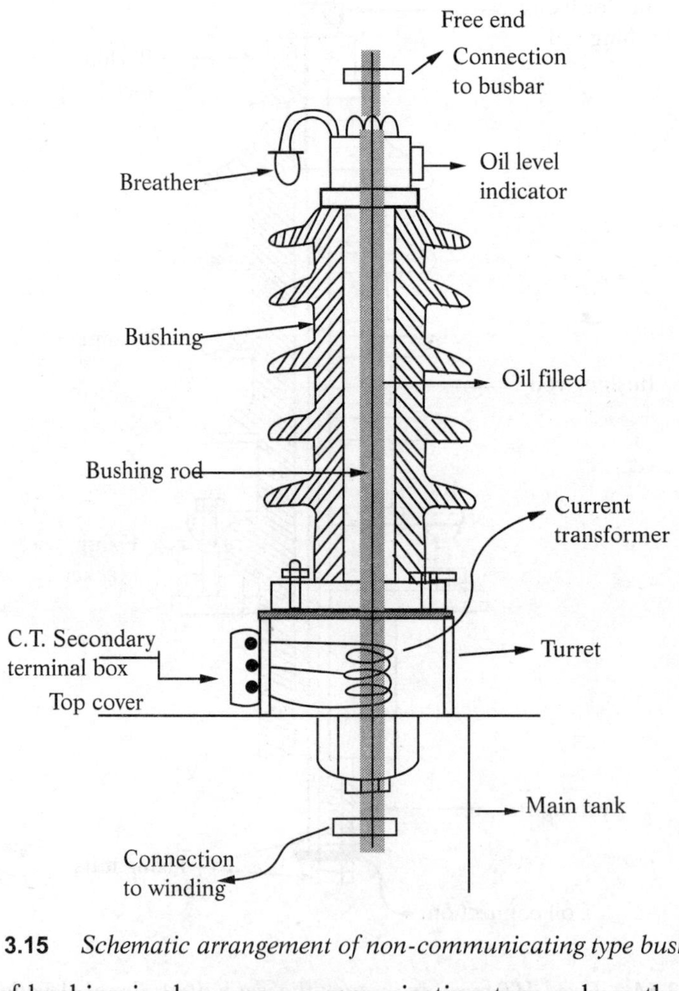

Fig. 3.15 *Schematic arrangement of non-communicating type bushing*

The third type of bushing is also a non-communicating type, where the bushing rod is covered with concentric cylindrical capacitors using paper insulation and metal foil. These capacitors are properly graded and impregnated with transformer oil. It is a hermetically sealed and no maintenance is required except for the cleaning and checking of the insulation and tanδ values.

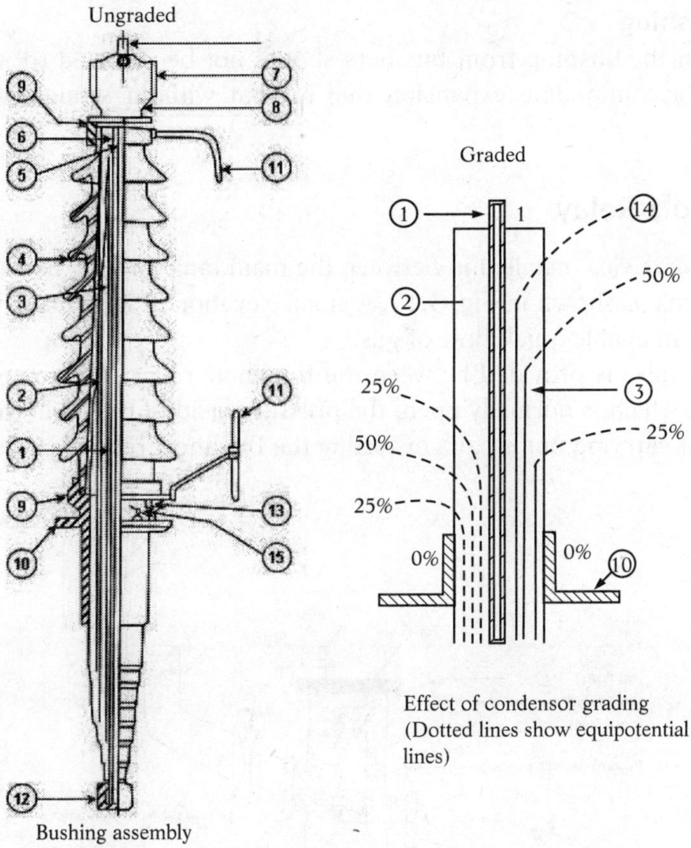

1 Central metallic tube
2 Condenser core
3 Conducting layers
4 Porcelain insulator
5 Cable thimble
6 Oil
7 Oil level gauge
8 Terminal gap
9 Gasket
10 Mounting flange
11 Arcing horn
12 Stress shield
13 Oil drain plug
14 Top terminal
15 Test tap

Fig. 3.16 *Condenser bushing components and effect of condenser grading*

The main advantage of this type of bushing is that it has high insulation with a considerably reduced size. Figure 3.16 shows the details of condenser bushing.

Turrets for housing CTs

The power transformer requires current transformers to be mounted on either voltage-level sides for protection and metering. The appropriate place for mounting these CTs is the bushings. Hence, a turret projects from the main tank or top cover on which the bushings are mounted. While the bushing connection rod is the bar primary, the wound secondary is housed inside the turret, surrounded by oil, and its secondary terminals are brought out to a terminal box for use.

The arrangement is shown in the Fig. 3.15.

Clamping in bushing

Fixed clamping in the bushing from bus bars should not be resorted to. Only free end clamps that can accommodate expansion due to heat without straining the bushings should be made.

3.1.9 Buchholz Relay

This is a protective device installed in between the main tank and the conservator on the connecting pipe link as shown in Fig. 3.17. A small elevation of 2° is usually given on the pipe axis in order to enable quick flow of gas.

A flow control valve is provided between the Buchholz relay and the conservator. By closing this valve, which is normally open, the pressure head of the oil in the conservator can be blocked for carrying out repairs or testing the Buchholz relays actions by pumping air.

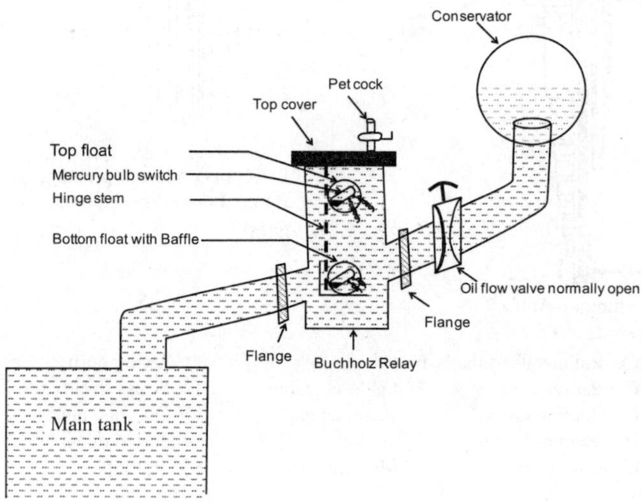

Fig. 3.17

The Buchholz relay chamber consists of a small metallic container, one side with a viewer glass graduated with volumetric measurement to note the gas collected, with scale reading indicated from top to bottom. The other side is closed with a Hylum sheet in which leads are connected. There is a top cover placed over the container and tightened with gasket. The top of the top cover has a pet cock to release the gas or air collected, which is also used for testing the Buchholz relay. Two floats are suspended through a metal rod and placed in the appropriate positions with the help of a hinge connected to the rod. The top float is placed just below the top cover and the bottom float is placed on the line of oil flow from the main tank to the conservator.

These floats will fall down on the hinge if there is no oil. When there is oil, these float due to buoyancy of the oil. Mercury contacts are placed on the side of each float with their terminals and leads as shown. The leads are brought out through bolted terminals on the side covered with the Hylum sheet. The mercury contacts are placed in such a position that when the floats are floating, the contacts are not bridged by mercury. But when the buoyancy is lost and the float falls down on the hinge, the mercury contact comes to a position with the mercury bridging the contacts. There is sufficient depth of the container from bottom float so that sludges, if any, collects at the bottom. The protection features exercised by the Buchholz relay is discussed in Section 4.1.5.

3.1.10 Explosion Vent

If any severe fault occurs on the transformer, heavy pressure of oil and gas is developed inside the transformer. If this pressure is not *vented out,* it will result in the bursting of the main tank. Hence, for proper venting out of the pressure, a device called *explosion vent* is provided. Its arrangement and description is shown in Fig. 3.18.

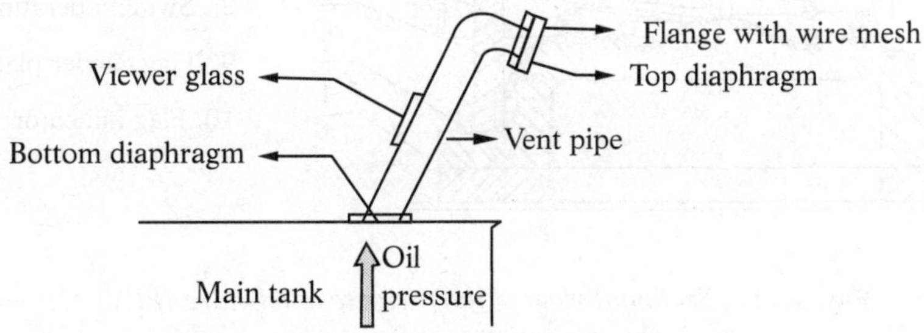

Fig. 3.18 *Explosion vent*

The vent pipe as shown in the figure is placed on the top cover with a *swan neck* top to enable the spurting of oil to fall away from transformer tank. The vent pipe has a top *diaphragm* protected with a wire mesh and a bottom diaphragm at the bottom of the pipe. Providing a bottom diaphragm in a power transformer is a must if a Buchholz relay is installed in order to ensure effective action of the Buchholz relay.

These diaphragms are of the *shattering type* such that when venting out pressure, these diaphragm are shattered to pieces for allowing the pressure of oil and gas to be vented out. Hence, usually 4 mm thick Hylumn sheets are used as diaphragms.

The top diaphragm, if broken due to age, is visible to the eye. But the bottom diaphragm, which is subjected to pressure due to head of the oil, may puncture or break in the course time due to ageing. Hence, a viewer glass is provided on the vent pipe, and if oil is seen on this viewer, it is a confirmation that the bottom diaphragm is broken and will have to be replaced at the earliest. The top diaphragm, if broken, has to be replaced immediately as well; otherwise, air with moisture will directly enter into the transformer. An equaliser pipe is also attached with the top of the vent, to be connected to the conservator.

3.1.11 Pressure Relief Valve

This is an alternative device to the explosion vent. It has a non-shattering type of arrangement as shown in Fig. 3.19.

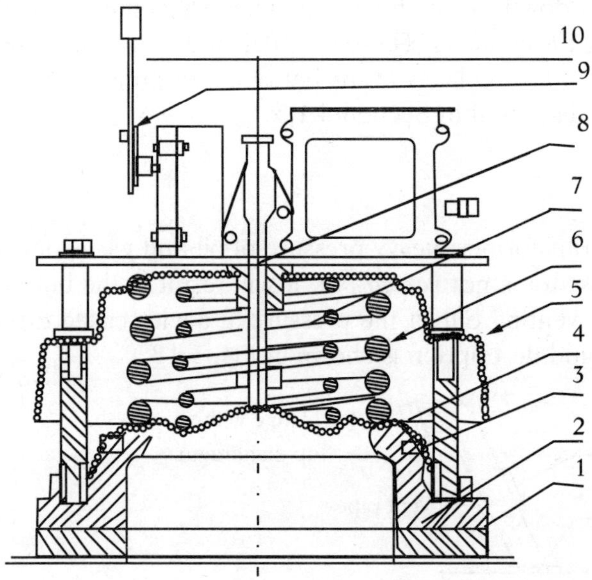

1. Base
2. Gasket ring
3. O Ring
4. Diaphragm
5. Cover
6. Spring
7. Rod retaining spring
8. Switch operating rod
9. Flag carrier plate
10. Flag indicator

Fig. 3.19 *Sectional view of the pressure relief valve (PRV)*

A steel diaphragm in contact with an O ring is pressed by a spring. The spring pressure is more than the pressure due to the head of oil. But if a heavy pressure is developed due to a fault, the spring is compressed and a port is opened by the lifting of the diaphragm and oil is spurted out. After the release of pressure, the spring takes over and closes the port.

Due to this action, a mechanical flag indicator falls down and connects to the protective circuit for isolating the transformer. The indicator should be brought back mechanically to its original position to cut off the trip circuit.

3.1.12 Equaliser Pipe Arrangement

All projected portions on the transformer tank, such as the turrets and the vent pipe, should be linked with a 12 mm pipe and connected to the top of the conservator. This will ensure that no air trap is developed inside the projected portions. If air is trapped inside the projected portions, it will push down the oil and cause the insulation to break down. Hence, this equalising pipe arrangement is an important protective device.

3.1.13 Air Release Cock or Plug

An additional device is provided on the top cover, called the air release cock or plug. This cock is to be opened and air, if any, has to be released periodically. The oil comes out after the release of air. The air release plug has a hole inside. By unscrewing the plug, oil comes out of the hole after the release of air. This arrangement is shown in Fig. 3.20.

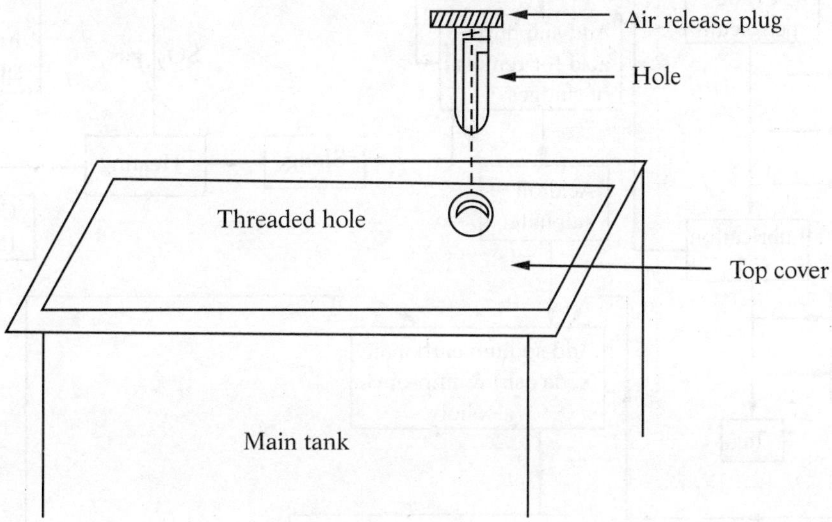

Fig. 3.20 *Air release plug*

3.2 Use of Oil in Transformers

3.2.1 Introduction

The transformer oil, widely used in electrical applications, is manufactured from the lubrication oil supplied as base stock for this manufacture, obtained during the refining process of crude petroleum. The manufactured oil should conform to the electrical and physical standards required for use in transformers of different voltage classes as an insulant and coolant. The flow diagram of the manufacturing process is given in Annexure I.

3.2.2 Standards for New Unused Oil

The electrical and physical standards required for newly manufactured transformer oil is given in Annexure II.

Annexure I: Flow diagram of manufacturing process

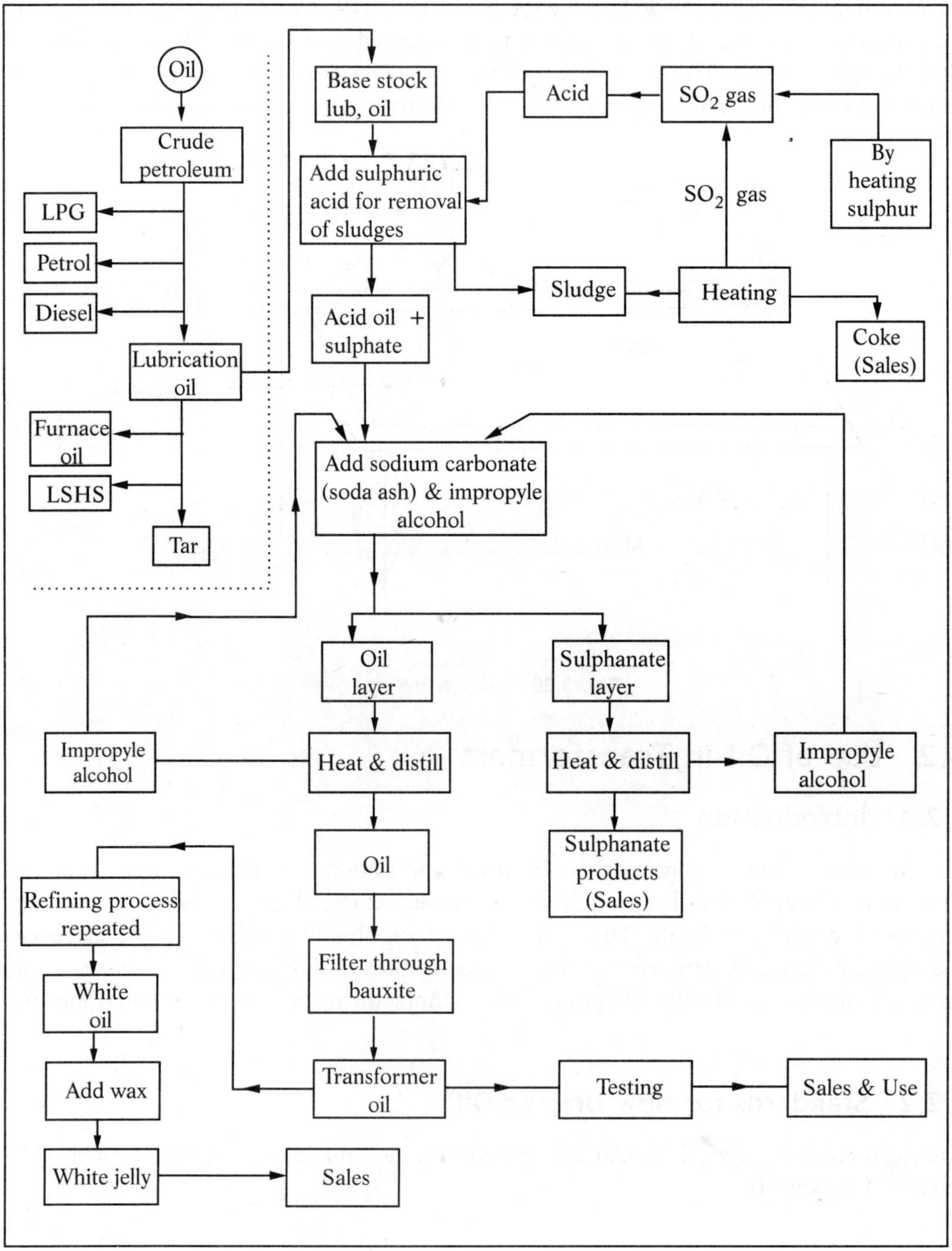

Annexure II: Properties exhibited by new transformer oil

As coolant

The physical properties to be exhibited to function as coolant are:

A	Density	:	0.89 at 27°C
B	Viscosity (Max)	:	27 Centi-stoke at 27°C
C	Flash point (Min)	:	130°C
D	Pour point	:	−6°C
E	Corrosive effect	:	Non corrosive

As insulant

The electrical properties to be exhibited to function as an insulant are:

A	Visual appearance	:	Clear, transparent, free from suspended impurities
B	Dielectric strength (BDV) (Min)	:	(i) As received – 30 kV RmS (2.5 mm gap) (ii) After filtration – 60 kV RmS (2.5 mm gap)
C	Water content	:	30 ppm
D	Resistivity i) At 90°C (min) ii) At 27°C (min)	 : :	 35×10^{12} ohm-cm 500×10^{12} ohm-cm
E	Total acidity (Max) (Neutralisation value)	:	0.03 mg of KOH/gm of oil
F	Inter facial tension(IFT) (min)	:	Not less than 0.04 N/m (940 dynes/cm)
G	Tanδ test (DDF) (max)	:	0.003 at 90°C
H	Ageing test (96 hrs. at 115°C)	:	(1) Resistivity -0.2×10^{12} ohm-cm at 90°C (min) (2) tan δ at 90°C –0.02% max (3) Total acidity –0.05 mg KOH (4) Total sludge –0.05% by weight

3.2.3 Use of Transformer Oil

Transformer oil is generally used in transformers and circuit breakers. The use of oil in circuit breakers is for insulation and arc quenching purposes. Hence, it is subjected to heavy carbonisation and gets contaminated quickly and frequently. The oil has to be, therefore, replaced and then reclaimed for the next use. Frequent reclamation through heating process makes the use of reclaimed oil limited to circuit breakers alone.

In transformers, only new transformer oil is used in the first time use or for top up. Oil is used as an insulant for transformers of various voltage ratings, from LT voltage to EHV voltages up to 400 kV. It is also used as a coolant to dissipate the heat produced in the transformer due to power loss.

As an insulant

As an insulant, the transformer oil must have the specific electrical properties as defined by various tests conducted periodically during the working period of the transformer. The permissible values in practice are defined separately along with the interpretation of the test results. It may be seen that these permissible values differ from that of the new oil, allowing for some deviation along the working period. It may also be noted that the dielectric strength of transformer oil (BDV) for various voltages are generally the same, since increased clearances are adopted for higher voltage classes.

As a coolant

The heat produced in the transformer core and the HV and LV windings during the working period of the transformer must be carried away for dissipation. The transformer oil serves this purpose by carrying away the heat to the main body and cooler tubes by convection process. The heat is dissipated to the atmosphere through the main tank and cooler tubes by conduction and radiation.

The basic arrangement for this cooling is the oil's natural circulation, a process that naturally occurs due to the tendency of hot oil to rise and circulate. If required further, additional cooling such as (i) air forced using fans and (ii) oil forced using pumps is resorted to for effective cooling.

The physical properties of the oil to function as coolant are:

(a) The viscosity of the oil should be such as to allow free flow at higher temperature and to carry heat but to return back to normal when the temperature falls.

(b) The oil should be non corrosive so that it does not react with metallic enclosures, windings and insulating materials.

(c) The density should be low so as to allow higher density contaminants like water to settle down at the bottom.

3.2.4 Possible Contamination of Oil

Transformer oil is always subjected to heating around 80°C. The heat, along with the atmospheric air maintained in the conservator, may gradually get contaminated. Unusual electrical and magnetic stresses due to adverse field conditions such as arcing and failures may affect the oil.

Thus, the oil may get contaminated gradually due to electrical and thermal stresses as detailed below.

(i) The heat, in conjunction with oxygen in the air, produces acidity in the oil due to which the oil becomes corrosive, affecting the insulation materials, metallic parts and winding conductors. Also, there is sludge formation that affects heat dissipation adversely.

(ii) The entry of moisture into the transformer pollutes the transformer oil and it finally gets absorbed in the windings resulting in deterioration of the insulation.

(iii) Suspended particles like dust entering the transformer gets absorbed in the oil leading to its contamination

(iv) Any hot spots developed in soldered or braced joints will result in a rise in temperature at the hot spot, resulting in contamination of the oil.

(v) Any arcing or sparking occurring due to partial or fully broken joints in current carrying wires results in high arcing temperature around 500°C, which contaminates the oil with suspended carbon particles.

(vi) Circulating current between the live and earthed parts due to stray voltages or partial discharges in insulations may cause local heating (without power follow through).

(vii) Higher power-follow current due to overload or short circuit, causing thermal and mechanical stresses

It may be seen that all the defects occurring during the working condition of the transformer will contaminate the oil. Depending upon the degree and nature of contamination, the performance of the transformer is affected.

3.2.5 Oil Test for Contamination and Interpretation

The following tests are generally conducted for the working condition of the transformer oil and thereby the condition of the transformer is ascertained.

Place of sampling

It is in practice to take oil samples in the following locations of a transformer:

(i) *Bottom sampling valve*: The moisture and sludges will settle down at the bottom and hence it is an ideal location for sampling.

(ii) *Top sampling valve*: The contamination is carried up to top cover along the movement of the hot oil. Hence, a sampling at the top is useful. Also, a comparison between the bottom and the top sampling can be made and analysed.

(iii) *At the conservator*: This is the place where the effect of oxidation with hot oil corrodes metallic parts inside the conservator above the hot oil and also where oil comes in contact with the atmosphere, providing direct exposure to moisture and suspended particles.

(iv) *On load tap changer*: Frequent OLTC operations may result in the OLTC oil getting contaminated with carbon particles. Also, other contaminants like moisture and dust may also make an entry.

Oil sampling procedure

(i) Oil sampling to be done in clean, dry weather

(ii) Sampling valves to be cleaned before sampling

(iii) Preferably, amber coloured bottles are to be used and they should be kept cleaned and dried before sampling.

(iv) Some quantity of oil should be flushed out before filling the sample bottle.

(v) After sampling, bottles are to be corked tightly and sealed with sealing wax.

(vi) Testing the sample at site is preferable.

Tests conducted

Screening test

By pouring the oil in a glass tube or tumbler, the colour, presence of suspended particles and clarity can be observed and interpreted as follows.

Colour	Pale yellow	Good oil
	Bright	Marginally good
	Amber	Bad oil
	Brown	Very bad oil
Particles	Presence of particles	Impure. To be tested.
Clarity	Hazy (poor visibility)	Excess moisture. To be tested.

Crackle test (free moisture test)

The presence of heavy moisture can be detected by this test.
The procedure is as follows:

(i) Take a metal rod of about 12 mm diameter.

(ii) Heat this rod to a dull red hot with blow lamp.

(iii) Have the oil tested in a clean and dry jar.
(iv) The hot rod is to be inserted in the oil and stirred.
(v) Crackling sound indicates the presence of moisture.
(vi) If it contains no moisture, there is no crackling sound.

The other method of detection of moisture is by adding anhydrous copper sulphate to a sample of oil. The oil will take a blue tint if moisture is present.

Dielectric strength test - breakdown voltage (BDV)

The BDV test is conducted by taking an effective volume of 300 ml to 500 ml of the oil in a transparent jar having two spherical polished electrodes placed at the centre of the jar with a gap of 2.5 mm in between. A voltage is applied across these electrodes and increased gradually from zero. The kV value at which the oil breaks down between the electrodes is the BDV. This increase of voltage from zero can be made uniformly at the rate of 2 kV/second either manually through a variac or through a small synchronous motor. Once the breakdown occurs, the test circuit is automatically tripped.

The test is to be carried out six times, and after each BDV the oil is gently stirred between the electrodes by means of clean dry glass rod while ensuring that there are no bubbles. The average of the six tests is considered as the test value for the oil. The general arrangement for the test is schematically given in Fig. 3.21.

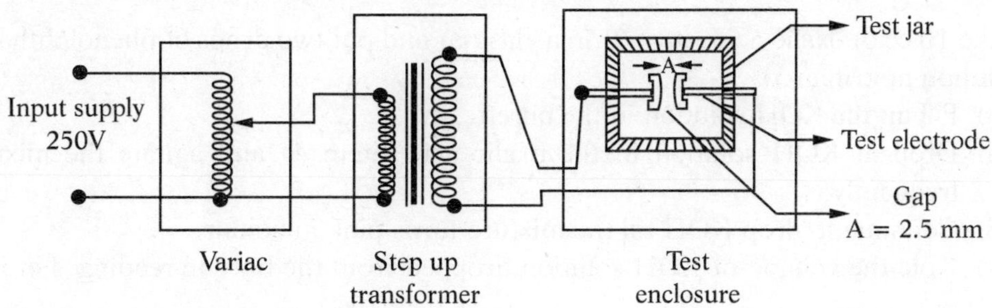

Fig. 3.21

The permissible test values of BDV are:

50 kV (min) for use in 220 kV and above

40 kV (min) for use in 110 kV

30 kV (min) for use in 66 kV and below

This test indicates electrical properties and it shows the ability of the oil to withstand electrical stresses due to high voltages low breakdown values occur because of suspended impurities and moisture.

Acidity test

This test exhibits the chemical properties of the oil. Refined insulating oils are free from mineral acids and alkalis, but they practically have natural organic substances which are weak acids. On oxidation during the working condition of the transformer, they form compounds which are acidic in nature. The organic substances present are referred to as organic acidity, whereas mineral acid and contamination present in the oil is referred to as inorganic acidity.

The total acidity test reveals the presence of both organic and inorganic components. In working condition, the organic acidity is negligible. Hence, for practical purposes, the measure of total acidity is accounted for by inorganic acidity.

Test for total acidity

This is conducted using titration using physical balance and *chemical solutions*.

Preparation of KOH solution

(1) Weigh 0.561 gm of KOH (potassium hydroxide) crystals and dissolve it in 100 cc of distilled water.
(2) Weigh 0.63 gm of oxalic crystals in 100 cc of distilled water.
(3) Weigh 0.05 gm of phenolphthalein powder and dissolve in 5 cc of distilled water.

Determination of strength of KOH

(4) Take 10 cc of oxalic acid (item 2) in a glass jar and put two drops of phenolphthalein solution in it(item 3).
(5) (i) Fill up the KOH solution in the burette.
(ii) Drop in KOH solution in the oxalic acid (item 4) and agitate the mixture frequently.
(iii) Continue to drop KOH till the mixture turns pink in colour.
(iv) Note the volume of KOH solution dropped from the burette reading. Let it be x.
(v) Strength of KOH $= 10/x$.

Finding of total acidity

(6) Weigh approximately 2 to 3 gm of oil in a small glass vessel. Add 2 to 3 cc of rectified spirit in the oil. Boil it with the help of a spirit lamp and shake the mixture.
(7) Add two to three drops of phenolphthalein solution (item 3) indicator to the oil and spirit mixture (item 6).
(8) Drop KOH solution (item 1) from the burette into the mixture (item 7) until pink colour is obtained and note the quantity of KOH solution dropped from the burette. Let it be y.
(9) Now total acidity $= 56.1 \times \frac{10}{x} \times \frac{y}{\text{Weight of oil}} \Rightarrow \ldots$ mgm of KOH/gm of oil.

Testing acidity range by colour indication

The above titration process is cumbersome and requires precision. Hence, a simple process by which a range of acidity is obtained is described below.

Acidity test by colour chart (source: CPRI, Bangalore)

Contents: The portable test kit box contains the following:

1. Two polythene bottles containing 100 ml each of ethyl alcohol and sodium carbonate solution of said concentration
2. An indicator bottle containing universal indicator, e.g., as supplied by M/s British Drug House Ltd.
3. Four clean glass test tubes
4. Three graduated droppers which serve as pipettes
5. Colour chart calibrated with neutralisation number values
6. Instruction booklet

Test procedure

1. Take 1 ml of insulating oil (approx 1 gm).
2. Pipette this quantity into a clean dry test tube.
3. Add 1.0 ml of rectified spirit (ethyl alcohol).
4. Shake the mixture gently.
5. 0.0085N solution of sodium carbonate of 1.0 ml is taken and added.
6. Shake test tube again gently.
7. Add 5 drops of universal indicator.
8. The resulting mixture develops a colour depending upon the pH value of the mixture.
9. Verify this colour with the provided colour chart having 10 colours for various values of acidity from 0 to 1 at intervals of 0.1.
10. Note down the *acidity value* of oil under test.

Preparation of 0.0085N sodium carbonate solution

Dissolve 0.1183 gm of anhydrous sodium carbonate in 250 ml of distilled water. One ml of this solution has just enough alkali to neutralize the acids extracted from the oil with acid number 0.5. Solution thus prepared has a life of 3 years, after which it should be discarded and a fresh one prepared.

Precautions to be taken during test

1. Test tubes and pipettes used should be absolutely clean and dry.
2. It is best to wash with soap and hot water followed by a final rinse with distilled water.
3. Use of chromic acid for cleaning is recommended wherever possible, but traces of acid to be completely washed off before drying.

4. Rectified spirit used should be of good quality. Commercial spirit should not be used. Ethyl alcohol AR grade is recommended.
5. During test the tube should be gently shaken using a thumb to close the test tube. Vigorous shaking may lead to contamination of the solution.
6. Same pipette can be used for a number of oils, provided it is rinsed with oil to be tested before pipetting.
7. Accurate acidity value of oil can be obtained by titration process only.

Permissible values for acidity

Safe value	Less than or equal to 0.5 mg of KOH per gram of oil
Observation value	Between 0.5 to 0.7 mg of KOH per gram of oil
Upper limit value	1.0 mg of KOH per gram of oil

Generally, for acidity value reaching 0.7 mg of KOH per gm of oil, there is rapid increase of acidity. Hence, it should be watched very frequently. When the value reaches 1 mg, the oil should be replaced with new oil after draining the high acidity oil completely from the transformer and flushing it with the new oil.

While in service, the transformer oil deteriorates gradually through oxidation with the resultant formation of sludge and other oxidation products. The effect of this acid formation is

(a) It promotes emulsification of water with oil and hence contributes to lowered dielectric strength.
(b) Peroxides formed will seriously reduce the mechanical strength of fabric insulation.
(c) It causes rusting of iron materials inside the tank above the oil level.
(d) It attacks the windings.

Hence, the acidity test is an important test to be made on all transformers, especially those subjected to heavy load and temperature. To prevent transformer failure, any high rate of increase in acidity should be investigated and remedial measures resorted to.

Other tests

The other tests that are carried out on the oil of the transformers in service are indicated below with their permissible values and interpretation. Modern testing instruments are available for these tests.

(a) *Moisture content* (physical property of oil)
The permissible moisture content is

(i) 25 ppm for voltages above 145 kV.

(ii) 35 ppm for voltages below 145 kV.

Higher content of moisture will affect the dielectric strength.

(b) *Flash point* (physical property of oil)

It is the temperature at which oil gives out so much vapour that this vapour, mixed with air, forms an ignitable mixture and gives momentary flash on application of flame under prescribed conditions.

A maximum flash point is specified in order to prevent the risk of fire that might result by accidental ignition. For all voltage classes transformers, this would be more than 125°C.

(c) *Resistivity test* (electrical properties)

The permissible value is 0.1×10^{12} ohm cm of the minimum value at 90° C. This is the resistance between the opposite faces of a centimetre cube of liquid insulation. It gives a measure of the presence of conducting contaminants, indicated by a low value of resisitivity.

(d) *Inter-facial tension test* (physical properties)

The permissible value is not less than 18 dynes/cm for all voltages. This is a measure of the molecular attractive force between the oil and water molecules at their interface. This test provides a means of detecting soluble polar contaminants and products of deterioration, which decrease the molecular attractive force between oil and water. This gives an indication of sludging.

(e) *Dielectric dissipation factor* (DDF) ($\tan \delta$) (electrical property)

This is numerically equal to the sine of the loss angle δ (at a low value of δ, $\sin \delta$ can be taken to be approximately equal to $\tan \delta$) and hence can be considered a function of the impurities present and the pure capacitive aspect of the oil. This is indicated Fig. 3.22.

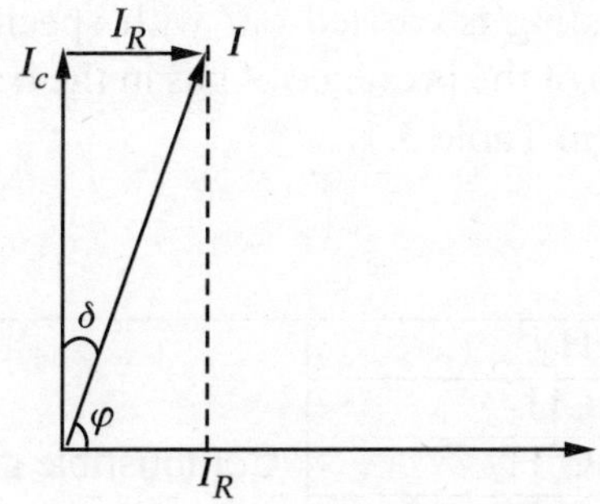

φ = Power factor angle
I = Total current
I_c = Pure capacitor current
I_R = Impurities current
δ = Loss angle
$\frac{I_R}{I_C} = \tan \delta = \sin \delta$

Fig. 3.22

The value of $\tan\delta$ should be as minimum as possible. In the working condition, the permissible values for the oil are:

At 90°C,

Max 0.005 – Good

Greater than 0.005 – Investigation to be done

Greater than 0.02 – Corrective measures should be taken

A high value of DDF is an indication of the presence of power loss contaminants or deterioration products like water, oxidation products, soluble varnishes and resins.

(f) *Gas analysis tests*

The description about the tests is given in Section 3.2.6.

3.2.6 Analysis of Gas in Transformer Oil

In the working condition of the transformer, the oil is always subjected to heat and there is a need for the evaluation of gas which can be classifies as (i) dissolved gas or (ii) liberated gas.

Arcing or sparking will result in gas being liberated from the oil, while hot spots, due to improper jointing, etc., will also produce gas but it gets dissolved in the oil itself. Valuable information about the fault development or developed fault can be ascertained by testing and analysing these gases and taking remedial measures to prevent failure, thereby increasing the life expectancy of the transformer.

Dissolved gas analysis

As a basic record, the presence of various gases in the oil during commissioning of the transformer may be ascertained. Further testing is carried out with specific periodicity during the working period to reveal the status of the presence of gas in the transformer oil.

The key gases present in the oil are given in Table 3.1.

Table 3.1

1	Hydrogen	–	H_2	Combustible gases
2	Methane	–	CH_4	
3	Acetylene	–	C_2H_2	
4	Ethylene	–	C_2H_4	
5	Ethane	–	C_2H_6	
6	Carbon monoxide	–	CO	
7	Carbon dioxide	–	CO_2	Non-combustible gases

The presence of these gases is detected by gas chromatography.

The gas chromatograph consists of:
(a) sample inlet system
(b) an inert carrier gas (mobile phase)
(c) a chromatographic column (stationary phase) and
(d) detection and recorder.

The procurement of a representative sample of oil from a transformer is important.

(a) The sample should be collected and transported in such a way that gases dissolved in the oil are not subjected to any change
(b) Amber coloured, dry and clean bottles should be completely filled up with sample, without any air column and completely sealed.
(c) An ideal container is a vacuum tight stainless steel bottle fitted with valve and gas cock.
(d) The interval between sampling and testing should be minimised.

There are two methods of analysis, namely key gas method and ratio method.

Key gas method
This method ascertains the value of gas available and the increase of their presence in the working condition and analyses them. The permissible values of key gases for a healthy transformer for different years of working condition are given in Table 3.2.

Table 3.2

Sl No.	Key gas	CHE Name	Quantity in ppm		
			Less than 4 years of service	4 - 10 years of service	Above ten years of service
1	Hydrogen	H_2	100–150	200–300	200–300
2	Methane	CH_4	50–70	100–150	200–300
3	Acetylene	C_2H_2	20–30	30–50	100–150
4	Ethylene	C_2H_4	100–150	150–200	200–400
5	Ethane	C_2H_6	30–50	100–150	800–1000
6	Carbon Monoxide	CO	200–300	400–500	600–700
7	Carbon Dioxide	CO_2	3000–3500	4000–5000	9000–12000

The recommended gas levels can be ascertained by using the formulas given in Table 3.3.

Table 3.3

n = Number of years of service; TCG = Total combustible gas

Sl No.	Key gas	CHE Name	Permitted concentration	Type of gas
1	Hydrogen	H_2	$< 20 \times n + 50$ ppm	Combustible TCG = $< 105 \times n + 710$
2	Methane	CH_4	$< 20 \times n + 50$ ppm	
3	Acetylene	C_2H_2	< 10 ppm	
4	Ethylene	C_2H_4	$< 20 \times n + 50$ ppm	
5	Ethane	C_2H_6	$< 20 \times n + 50$ ppm	
6	Carbon Monoxide	CO	$< 25 \times n + 500$ ppm	
7	Carbon Dioxide	CO_2	$< 100 \times n + 1500$ ppm	Non-combustible

The interpretation of the presence of TCG is given in Table 3.4 (Westing House guidelines).

Table 3.4

1	0 to 500 ppm	Represents normal ageing. Analyse again in 6 months.
2	501 to 1200 ppm	Indicates decomposition in excess of normal ageing. Analyse in 3 months.
3	1201 to 2500 ppm	Indicates decomposition quite in excess of normal; analysis required in one month.
4	2501 ppm and above	Make weekly analysis to determine rate of decomposition (contact manufacturer).

Note: The key gas method is more useful in analysing the *developed fault* than for analysing the *developing fault*.

Roger's ratio method

This method describes various combinations of ratio of gases and their interpretation, and is more useful in detecting pre-fault condition. In case of severe fault, both Key gas and Rogers methods can be employed to ascertain the conditions of severity. The ratio combinations used in this method are as follows:

(i) CH_4 / H_2 ii) C_2H_6 / CH_4

(iii) C_2H_4 / C_2H_6 (iv) C_2H_2 / C_2H_4

A ratio less than 1 is defined as zero [0] and more than 1 is defined as one [1]. The interpretation of ratio is given in Table 3.5.

Table 3.5

Sl. No.	$\frac{CH_4}{H_2}$	$\frac{C_2H_6}{CH_4}$	$\frac{C_2H_4}{C_2H_6}$	$\frac{C_2H_2}{C_2H_4}$	Diagnosis
1	0	0	0	0	If CH_4/H_2 is less than 1 - partial discharge; otherwise normal deterioration
2	1	0	0	0	Slight overheating below 150°C
3	1	1	0	0	Slight overheating below 150°C - 200°C
4	0	1	0	0	Slight overheating below 200°C - 300°C
5	0	0	1	0	General conductor overheating
6	1	0	1	0	Circulating current or overheated joints
7	0	0	0	1	Flashover without power follow current
8	0	1	0	1	Tap changer selector breaking current
9	0	0	1	1	Arc with power follow through current or persistent sparking

Analysis of liberated gas in oil

(i) Buchholz relay is a gas actuated relay, actuated by liberated gas.
(ii) Liberation of gas means there is a fault developed.
(iii) Incipient fault actuates top float of Buchholz relay.
(iv) Heavy fault actuates bottom float.
(v) Test conducted in liberated gas:

  (a) Combustibility test: The gas is collected in a glass tube; a lighted matchstick will reveal combustibility. If not combustible, it is mere air.
  (b) Colour and quantity of gas: If the collection has no colour, it is merely air.
  (c) Chemical test: The gas is passed through two chemical solutions, namely: (1) dissolving 5 gm of silver nitrate ($AgNO_3$) in 100 cc of distilled water and (2) dissolving 5 gm of silver nitrate ($AgNO_3$) in 100 cc of watery ammonia solution.

Analysis:

Solution 1	:	Gas causes white precipitate which turns brown under the influence of light	Decomposition of oil by occurrence of flashover.
Solution 2	:	Dark brown precipitate is caused	Solid insulating materials decomposed, producing carbon monoxide (inter-turn short on the winding)

3.2.7 Reclamation of Oil Using Filters

The contaminated oil can be reclaimed using oil filters. The acidic contamination can be removed only by original refining process and hence not possible while in usage. However, other contaminations like suspended particles and moisture can be removed by circulation through filters. Generally, the following filters are used:

(1) Blotter press filter
(2) Streamline filter
(3) Centrifuge filter

The blotter press filter is used only for removing suspended impurities, whereas the other two filters are used for removing suspended impurities and moisture also. Generally, filtration through blotter press filter precedes the other filters.

Blotter press filter

The blotter press filter has many filter plates of adequate thicknesses comprising blotter papers, one or two (fixture mat), inserted between two plates and kept completely tightened as a filter array with two end plates. There is an inlet hole and an outlet hole through which the oil is forced in by a pump. The flowing oil has to reach the outlet only through the filter paper. The suspended particles are thereby filtered and thus the oil is purified. The papers must be changed when they become clogged. Clogging is indicated by the decrease in rate of flow for a given pressure drop across the filter.

The working requirement of the filter papers is:

(i) The papers should absorb water also from the oil and hence it should be kept in a dry place until taken for use.
(ii) The papers should not release fibres into the oil.
(iii) Filtering is done only at atmospheric temperature.

The figure of a typical blotter press filter is shown Fig.3.23.

Centrifuge filter

Since oil and water have different densities, filtering by centrifugal action throws out the water from the oil. Here also the use of filter is at atmospheric temperature.

Steamline filter

Removal of water by evaporation is done through this filter. Water will evaporate at 100°C, but if the oil is heated to such a high temperature, acidity becomes high through oxidation. Hence, this treatment is done under vacuum, where water will evaporate at 60°C. Therefore, it is enough if the oil is heated to around 60°C and the vacuum is maintained with a negative pressure of 700 mm mercury. The filter pack will remove suspended contaminants. The schematic diagram of a streamline filter is shown in Fig. 3.24.

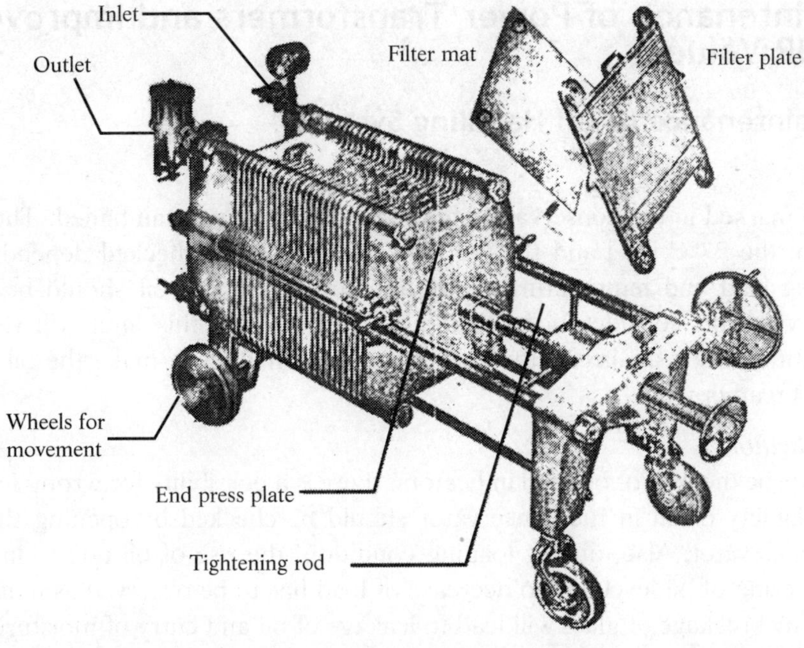

Fig. 3.23

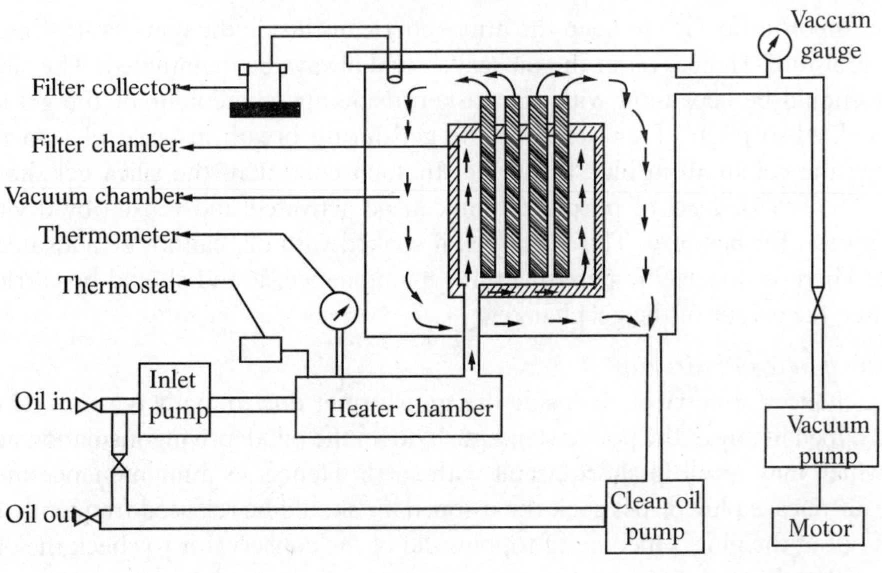

Fig. 3.24

3.3 Maintenance of Power Transformers and Improvement of IR Values

3.3.1 Maintenance of Oil Handling Systems

Conservator

The oil level marked in the conservator should be checked and maintained. The marked level between the 27°C level and the full load level has to be checked depending upon the prevailing load and temperature of the oil. Any top-up of oil should be made in accordance with this consideration. Topping up of oil over this limit will reduce the volume of atmospheric air inside the conservator and may even make the oil overflow through the breather pipe.

Oil level indicator

Whatever may be the type of oil level indicators, there is a possibility for wrong indication. Hence, availability of oil in the conservator should be checked by opening the top-up lid of the conservator. Also, during loading condition, the rise of oil due to increase of load and lowering of oil level due to decrease of load has to be observed as a day-to-day operation. Any breakage of glass will lead to leakage of oil and entry of moisture into the transformer. The broken glass should be replaced immediately.

Breather

This is an important device to keep the atmospheric air inside the conservator free from dust and moisture. Hence, oil in the oil seal should always be maintained. The silica gel container should be kept tight with the gasket inside and the colour of the gel should be observed. Absorption of moisture by the gel during breath-in cycle of transformer will change the colour from blue to purple. In such condition, the silica gel should be replaced. The released gel of purple colour can be activated and reused by drying the same in *hot sun* for next use. This silica gel, if soaked with oil, cannot be activated back for reuse. There is no specific periodicity for maintenance; it and should be carried out as and when the colour of the gel changes.

Maintenance to avoid air-trap

There is a chance for entry of air inside the transformer due to many reasons. Thus, air may get trapped in any of the pockets and push down the oil, depriving insulation at these pockets which may result in short circuit with earth. Hence, as a maintenance measure using the air release plug or pet cock the trapped air should be released frequently till the oil comes out of the plug. Opening of topping-lid of the conservator to check the oil level is also considered as an air release measure.

Maintenance of bushing

The bushings should be kept cleaned of dust as deposit of dust will affect the insulation through the creepage path. Except for hermetically sealed bushing and communicating type bushing, the other type, i.e., non-communicating type with separate oil should be maintained by testing its oil periodically and by topping up with new oil if oil level falls. Also, its breather and gauge glass should also be maintained. In case of condenser bushing, the value of the graded capacitance should also be periodically checked. The bushings should be ensured with the help of free end clamp connections to bus bars to prevent damage of bushings due to temperature stresses. This connection should be checked periodically.

Maintenance of main tank and cooling system

The cooling system plays an important role in loading the transformer up to its full capacity. Hence, effective maintenance should be assured. Fans and pumps operation are to be insured with spare fans and pump to takeover in case of emergency. Heat dissipation occurs not only in the cooling system but as radiation from metal tank as well. Hence, leakage of oil and absorption of dust in the tank and cooling pipe will also reduce the effectiveness of the cooling. Therefore, proper maintenance should be ensured to keep all the metal parts clean. Periodical observation of hot spot temperature through infrared thermometers will reveal hot spots in windings.

3.3.2 Testing of Oil Samples

The testing of oil samples from the transformer as described in Section 3.2 forms another activity of maintenance since it reveals the working condition of the transformer. The oil samples should be drawn from the following locations of the transformer.

(i) Bottom sampling valve
(ii) Top sampling valve
(iii) Conservator
(iv) OLTC

3.3.3 Measurement of Insulation Resistance Values (IR Values)

Periodical checking of the insulation resistance values (IR values) of the transformer and maintaining the same within the prescribed limits is the third activity of maintenance. This test is conducted not for the winding alone but for the whole transformer including oil and bushings.

The IR value is generally obtained using (i) 1000 volt megger or (ii) 2500 kV power operated megger by injecting the above specified DC voltage across the terminals to be tested. The 1000 volts megger will reveal only a value in the range of mega-ohms, and hence the exact value cannot be specified. But with a power operated megger, the exact IR value can be measured. Generally, a megger dial with 2000 MΩ range is comfortable to use.

The tests that can be conducted by the megger are (i) IR value test and (ii) continuity test for checking the interconnections.

(a) **Using a 1000 volt megger**

The IR value is to be tested only after disconnecting the earth link from the neutral as shown in Fig. 3.25.

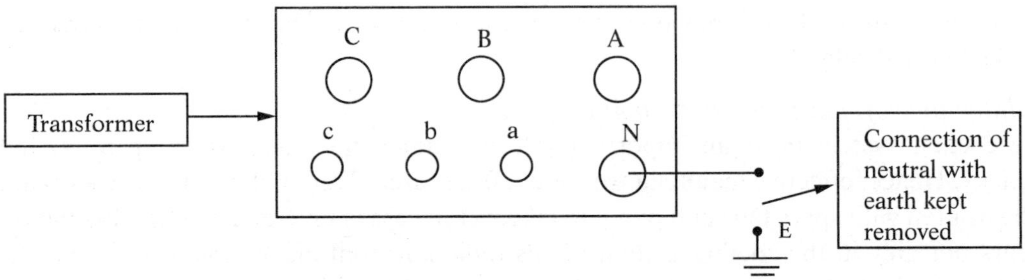

Fig. 3.25

Connecting the megger leads between any two terminals and taking measurement for IR value and continuity is recorded as follows:

IR value

HT to Earth	LT to Earth	HT to LT
A – E : ...... MΩ	a – E : ...... MΩ	A – a,b,c and N: ...... MΩ
B – E : ...... "	b – E : ...... "	B – a,b,c and N: ...... MΩ
C – E : ...... "	c – E : ...... "	C – a,b,c and N: ...... MΩ
	N – E : ...... MΩ	

Continuity test

HT		LT	
A - B B - C C - A	} 0 Ω	a - b, c and N b - a, c and N c - b, a and N	} 0 Ω

This test will reveal the general health of the transformer. The value should be recorded along with the temperature of the transformer at the time of the test. The temperature at which the measurement is taken is of prime importance since IR value is inversely proportional to temperature. Such measurement at 60°C temperature has to be considered as a reference value and should be 2 MΩ/kV. The recommended minimum IR values for specific voltages at a given temperature are given below if the reference value is not obtained. The variation of IR value is 2:1 for every 10° C variation inversely, and thus can be interpreted.

kV	IR values in MΩ at 30°C
0.44 kV	100 MΩ
Less than 6.6 kV	200 MΩ
6.6 kV to 22 kV	300 MΩ
33 kV	400 MΩ
66 kV and above	500 MΩ

After the test, the neutral should be linked with earth before putting the transformer into service.

(b) Polarization index

While measuring the insulation resistance value (IR value) of a transformer with a megger, a DC voltage V_{DC} is applied at the measuring terminals. A current I_{DC} flows. The IR value is shown in mega ohm as a measurement of V_{DC}/I_{DC}.

This I_{DC} is a total leakage current as seen by the megger and is defined by a combination of currents as follows:

$$I_{DC} = I_i + I_s + I_d + I_c,$$

where

I_i = constant conduction current through the insulation
I_s = constant surface leakage current
I_d = dielectric absorption current
I_c = capacitance charging current.

In the above combination, I_c and I_d flow for some duration of several seconds or minutes depending upon the size of the winding and capacitance of insulation. The value of I_c and I_d reduces with time and, after few seconds in small capacity distribution transformers and after a few minutes in large capacity EHV transformers, dies down to zero.

The remaining two components, I_i and I_s, are constant for a given applied DC voltage by the megger and hence are true indicators of the IR value of the transformer.

I_i increases if the insulation is in wet condition by absorption of moisture. I_d increases if dust particles are increased in the insulation. Since the meggering is done along with the bushings of transformers, wet insulation and dusty surfaces of bushings also increase these constant leakage currents and consequently lower the IR value. The die-down time for I_c and I_d may be taken as (i) 60 seconds for small capacity distribution transformers, and (ii) 10 minutes for high capacity of EHV transformers.

The magnitude of IR value alone is not taken into consideration. An additional characteristic called *polarization index* (PI value) is also considered as indicative of the degree or dryness of the transformer. PI value is the ratio of IR value measured in the presence of I_i and I_s alone to the I_R value measured in the presence of I_c and I_d also with I_i and I_s.

Hence, PI value for

$$\text{Distribution transformer} = \frac{\text{IR value measured at 60 seconds}}{\text{IR value measured at 15 seconds}}.$$

$$\text{EHV transformer} = \frac{\text{IR value measured at 10th minute}}{\text{IR value measured at 1 minute}}.$$

Permissible PI value for power transformers at 30°C ≥ 1.3.

If the value is less than 1.3 and close to unity, the condition of insulation is in doubt and in danger, warranting immediate attention.

3.3.4 Improvement of IR Value

Continuous monitoring of the insulation resistance value (IR Value) of transformer and keeping the same at the required level will help maintain the health of the power transformer. The entry of moisture and other suspended impurities into the transformer will lead to low IR values. In such cases, arrangement should be made to improve the IR value.

The transformer should not be kept idle. It should always be put in the heat run by connecting some load. Hence, it is customary to keep all transformers in the sub-station in parallel, even though the level of load is within the capacity of one transformer (instead of cutting out the transformer and keeping it as a spare without load). Keeping the transformer at atmospheric temperature without use may lead to the lowering of IR values.

The method adopted for improving the IR values are are listed below. They are called 'drying out' process and utmost care must be exercised during the process to avoid damage to the winding or the oil.

(i) On no account during the drying out process must the transformer be left without attendance by proper personnel.

(ii) The insulation testers (megger) and the thermometers used should be reliable.

(iii) The life of the insulation decreases with every drying process. Hence, during the process, the temperature should be kept at least 5°C less than the maximum stipulated temperature for the transformer.

Hot air circulation

For small distribution transformers the core-coil unit can be lifted and placed in a hot air chamber and dried. The chamber can be made to accommodate 3 or 4 such cores at a time. But in case of large power transformers, the core-coil unit has to be dried keeping it in its main tank itself. Hot air is pumped inside the tank by a blower and a heater through the bottom of the tank and allowed to escape through an opening on the top cover diagonally offset with the entry point of hot air. This arrangement is shown in Fig. 3.26.

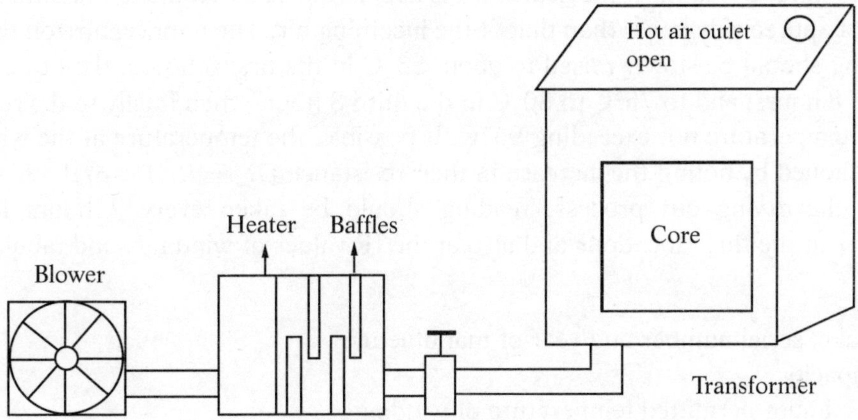

Fig. 3.26

The preliminary requirement for such circulation is that the oil in the main tank has to be completely drained out as possible. The oil in the cooling fins has also to be drained out and the header valves kept closed. The transformer tank should be completely covered with tarpaulin. This is to prevent loss of heat; it also helps to keep the whole of the interior tank at such a temperature as to prevent condensation. Thermometers can be fixed at the following places to record the temperature at the location.

(i) Inlet air pipe between the heaters and the transformer
(ii) The core and the windings
(iii) the air outlet point

Use of pencil thermometers may be avoided because if it breaks, mercury in the bulb will precipitate in the transformer windings. Hence, it is desirable to use dial type thermometers with metal bulbs and flexible capillary tubes.

The hot air is supplied by means of a blower through electric heaters, with proper control arrangement to regulate the temperature. The mouth of the blower where the cool air enters should be covered with one or two layers of cheese cloth to prevent dust from being drawn from the outside and blown into the windings. Strip heaters can be used instead of filament heaters to avoid naked flame. Baffles should be placed at the entry of hot air from the heater to the transformer to prevent direct passage of air with fire particles. If the blower has to stop, the heaters must also be automatically switched off to prevent overheating.

If possible, suitable baffles may be inserted between the core and tank inside so that heat blows uniformly through the ducts in the winding by closing the intervening surface as much as possible. Generally, the permitted winding temperature is 95°C and hence the maximum heat maintained should not exceed this limit during the process. To ensure uniform drying, the volume of the heated air is kept as such so that the temperature at the outlet is not appreciably lower than that of the incoming air. The temperature on the core and winding should be slowly raised to about 55°C in the first 8 hours; then to 65°C in the second 8 hours; and to 75°C to 80°C in the third 8 hours; then finally to the required maximum temperature not exceeding 95°C. If possible, the temperature at the windings may be reckoned by noting the increase in their resistance $[R_2 = R_1(1 + \alpha t)]$.

During the drying out process, reading should be taken every 2 hours for the temperature at the three locations and also at the IR values of windings and tabulated as follows:

(i) Make, serial number and year of manufacture
(ii) Capacity
(iii) Maximum permitted temperature of winding

Date & time	Temp at			IR Value Primary Phase - Earth			IR Value Secondary Phase - Earth		
	Inlet	Wdg	Outlet	A	B	C	A	B	C

As the temperature of the core and windings rises, the IR value will tend to fall, indicating release of moisture. This will eventually reach a steady value. The process should be continued till the value shows an upward trend. The upward trend indicates complete release of moisture. At this stage, the drying out process can be considered as complete and the circulation of hot air may be discontinued and the cooling cycle begun. During this cooling cycle also, the IR value has to be measured and recorded for every 5°C of lowering of temperature. The value at 60°C is an important deciding value.

This value at 60°C should not be below 2 megaohms per kV of the voltage rating of the transformer. The IR value varies with different makes and types of transformer during the process, but the lowering trend to the minimum and consequent rise as described earlier has to be watched during the drying out process. Should the IR value be high and not going down as the windings are heated up, it indicates the transformer is dry and does not contain moisture.

After the drying out process is completed, the core and windings should be flushed with hot, clean oil. The drained transformer oil before the drying out must be reconditioned to the required test value of 40 kV and this tested oil should be heated to 60°C and filled in the transformer through the filter.

Generally, three to four days of continuous circulation after attaining the maximum temperature of 95°C is carried out. If the transformer is not properly dried in this cycle, it may be given a cooling cycle by gradually bringing down the temperature to 40 to 45°C and then again raising the temperature and maintaining it constant for another three to four days. The cycle may be repeated till the transformer is completely dry.

Hot oil circulation with short circuit

The hot air circulation is the best method of carrying out the drying out process. In case of medium size power transformers and distribution transformers where the moisture content is low and IR value is not significantly low, this method of hot oil circulation with short circuit can be resorted to. In this process, the moisture in the windings and core are driven out by heating the winding not by actual loading but by loading through short circuit; oil circulation is also carried out simultaneously.

As already discussed, the percentage of impedance of the transformer is also called % impedance volts. It means that the maximum volts required to be fed on one side with the other side short circuited to drive the full load currents on the respective sides, and it is calculated as follows.

For example, if it is a 11 kV/433 V 100 kVA transformer with 5% impedance, 550 volts is the maximum of voltage to be applied on the 11 kV side (i.e. 5% of 11 kV) with the LT side short circuited to drive its full load current of 5.25 A / 133 A. This can be achieved by a special variac of suitable design to withstand the current being handled.

It is not necessary that the full load short circuit current need be maintained. With lesser voltage supplied, 60 to 70% of the full load current through short circuit is enough to be maintained so that the heat produced in the winding liberates the moisture. In this process the core and coil should be completely immersed in oil with sufficient space between the top oil and top cover. Since short circuit current is driven continuously, it is necessary that the core coil unit be kept immersed in oil. Thus, the driven out moisture from the winding is sent to the oil. The oil in turn is sucked through the filter wherein the moisture is separated.

The process of hot oil circulation with short circuit is shown in Fig. 3.27 schematically.

Just like in hot air circulation, the following has to be followed:

(i) The transformer has to be completely covered with tarpaulin to prevent heat dissipation.

(ii) If the transformer is fitted with radiator valves, the upper radiator header valve is only to be closed to prevent oil circulation, leaving the lower radiator valve open. Both upper and lower radiator valves should never be closed, since oil pressure caused by expansion will seriously damage the radiator valves.

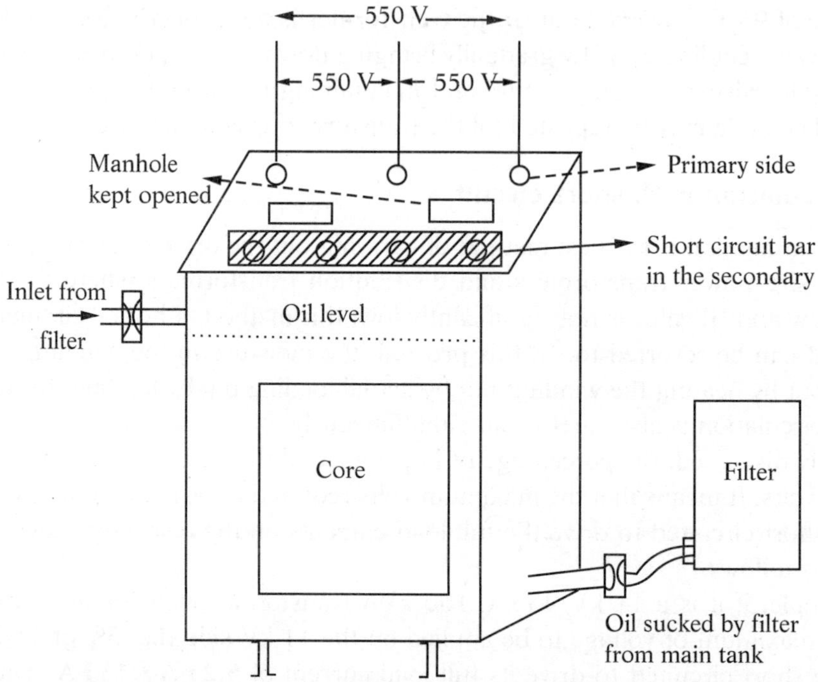

Fig. 3.27 *Process of hot oil circulation with short circuit*

(iii) Plenty of ventilation is to be provided in the air space above the oil to facilitate evaporation of moisture from the oil. The absence of such ventilation may result in moisture condensation under the side of the top cover, which will drop back into the oil. Keeping the manhole covers raised would normally provide adequate ventilation. Monitor to see that no condensation takes place during the process.

(iv) The temperature measurements has to be carried out for the oil at the top layer, and if possible, the winding temperature calculated from the value of the increase in resistance as discussed before.

(v) The temperature should be slowly increased from the atmospheric temperature to a value around 75°C. The filter is operated under vacuum, and hence the oil treatment will evaporate the moisture in the oil and remove it at the reduced temperature.

(vi) As already described, the temperature being kept constant, the insulation value is measured every two hours and the results recorded during the process. There will be a lowering trend, and it is necessary to keep up a steady value. The increasing trend indicates that the transformer has been completely dried. After registering the increasing trend for some hours, the short circuit and oil circulation are stopped. The IR value at each 5°C temperature reduction is recorded with the 60°C value as the criterion (greater than 2 MΩ per kV).

(vii) During each measurement of IR value, the short circuit supply should be switched off as safety measure and then the IR value taken and recorded.

(viii) The general period for this continuous circulation is 10 days. If sufficient improvement of IR value is not attained, only *hot air circulation* should be arranged for.

(ix) Continuous hot oil circulation through the filter will lead to a gradual deterioration of oil due to the continuous mechanical action and filtration under high temperature causing partial breakdown of the constituents of oil; it may also increase the acidity of oil. Hence, care should be taken check the oil quality as well during the process.

Drying out under vacuum

This method is to suck the water vapour through vacuum, liberated due to heating of core-coil unit due to short circuit method. It should be ensured that the transformer main tank has been manufactured with strong steel plates to withstand external air pressure resulting from the vacuum inside. The transformer has to be put under short circuit, as described earlier, with core-coil unit completely immersed in oil. The short circuit current should be near full load current by applying the maximum impedance voltage for quick drying, but should not exceed the rating since it may damage the windings. The transformer should be kept completely covered by tarpaulin and the radiator valves kept closed to prevent loss of heat.

A vacuum pump should be connected to the top of the transformer tank. The most convenient place is the pipe flange to which the breather unit is attached. The breather unit is removed and the flexible metallic tubing from the vacuum pump is connected to its flange. The diaphragm in the explosion vent should be removed and sealed with dummy plates through gaskets.

The temperature is slowly raised to 55°C in the first 8 hours from the start, to 65°C in the second 8 hours, and to the maximum temperature in the next 8 hours. After reaching the maximum temperature, the transformer should be disconnected from the source of power for short circuit and the oil drained out and vacuum applied immediately. The vacuum should be maintained at 700 mm of mercury, until temperature drops down to 50°C. After reaching 50°C, the vacuuming should be stopped and transformer oil at 60°C should be filled up through a filter. Then the short circuit is switched on. This cycle is repeated until the required IR value is obtained.

The temperature and insulation resistance of windings should be taken for every 2 hours and the process continued till the insulation value become steady and tends to rise. The interpretation of the megger reading is difficult as variation in temperature is involved in each cycle. Hence, a curve showing the variation of insulation resistance with time and temperature may be drawn (Fig. 3.28).

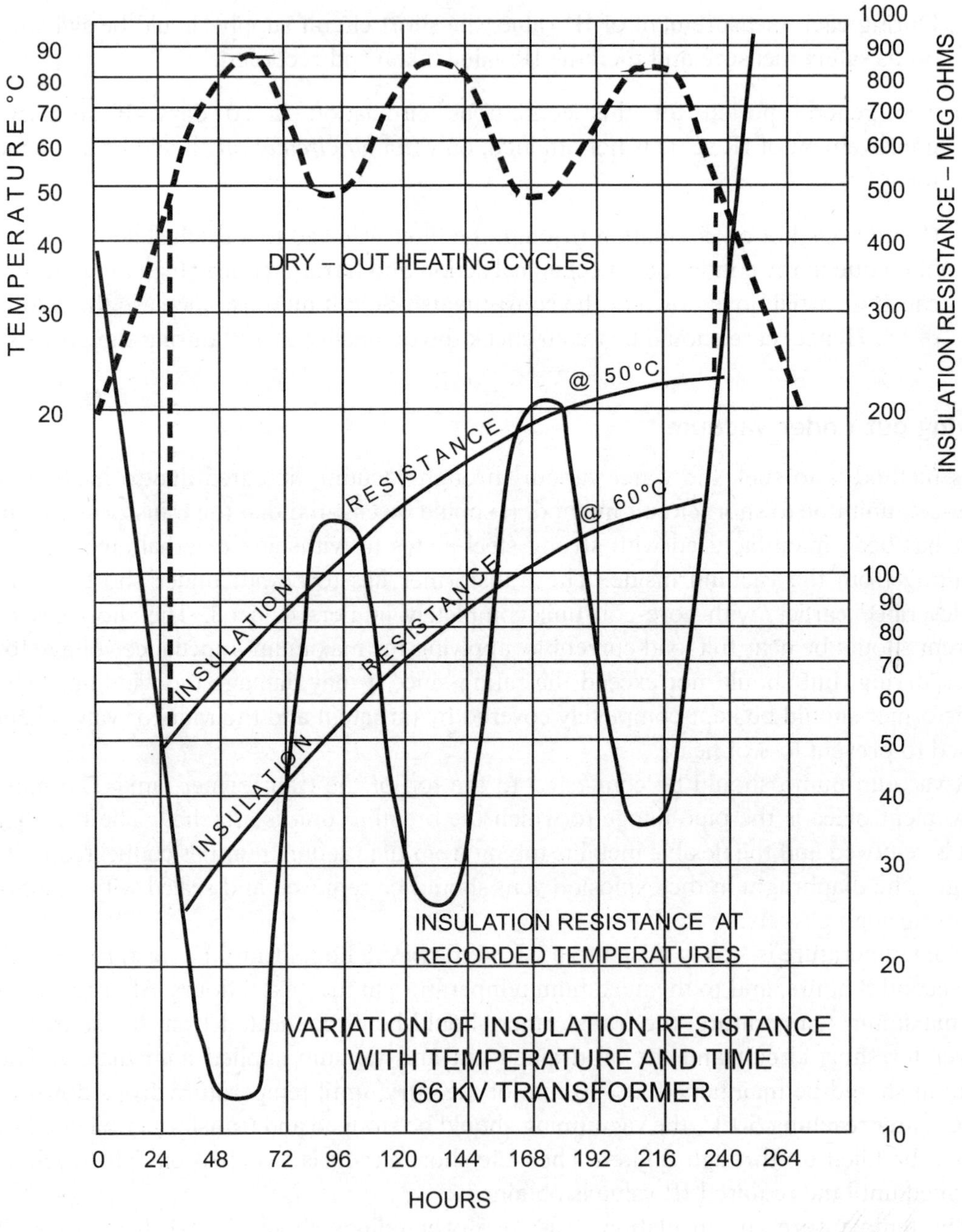

Fig. 3.28 *Insulation drying curve*

3.4 Troubleshooting Power and Distribution Transformers

Though the working conditions of the transformer had been described in detail in the previous sections, it is useful to list out the possible causes for troubleshooting and corresponding remedies (preventive and predictive). This has been done in Table 3.6.

Table 3.6

Trouble	**Cause**	**Remedy**
1. **Magnetic core** (a) Unusual hum	Loosening of core during transport or due to vibration during working condition or due to improper butting of limb and yoke laminations. Results in increased losses and magnetising current.	Ascertained by measuring no-load losses and magnetising current. Proper jointing of yoke and core stampings and tightening of loosened core bolts or strap bands.
(b) Unusual heating and abnormal rise of temperature	The insulation on both sides of lamination stampings would have deteriorated due to corrosion, resulting in heavy flow of eddy current causing abnormal temperature rise.	Corrosion is because of moisture or high acidity value of oil due to high temperature. Damaged laminations to be replaced or coated with fresh insulation and measures to be taken to avoid moisture entry and rise of acidity.
2. **Windings** (a) Failure of inter-turn or inter-layer insulation causing short circuit of turns, resulting in failure of coil stack.	(i) Continuous overload resulting in damage to insulation due to abnormal heating. (ii) Possible moisture entry causing low IR value or by acidic corrosion.	Loading has to be limited within its rated capacity and overloading avoided. Temperature increase to be kept within 70% of stability limit. Preventive maintenance on oil handling system and checking acidity value periodically.
	By conducting DC resistance test and magnetic balance test, the extent of failure is analysed and failed coil stacks replaced.	

<table>
<tr><td rowspan="2">(b) Development of hot spot due to broken joint, delta connections and bushing lead connections</td><td>Due to improper jointing or due to vibration</td><td>Effective soldering of joints should be made, housed inside thick fibre sleeves and properly fastened with structural arrangements</td></tr>
<tr><td colspan="2">Location identified by:
(i) Ascertaining the hot spot by infra-red thermometer
(ii) By DC resistance test, short circuit test or by Buchholz relay top float action</td></tr>
<tr><td>(c) Shattering of winding</td><td>Due to heavy short circuit forces acting on the windings, beyond the permitted duration - weak coil shatter</td><td>(i) Coils to be replaced with proper mechanical rigidity and electrical stress capability. Failed coils ascertained by physical examination

(ii) Proper protection schemes with circuit breaker or fuses ensure preventions

(iii) Proper maintenance of LV system</td></tr>
<tr><td>(d) End coil shattering</td><td>Entry of high voltage, high frequency surges into the winding</td><td>Function of surge diverters (LAs as locally called) and its effective earthing system checked and measures taken to set right the requirement</td></tr>
<tr><td>3. Oil
Sample not satisfactory from standard test results</td><td>Absorption of moisture and dust, increased acidity and dissolved gases due to heating, result in low IR value</td><td>(i) Transformer put on dry air heating and oil circulation for filtering, de-moistening and de-gassing

(ii) Overhaul if heavy sludges are noticed</td></tr>
</table>

4. **Bushings**		
(a) Undue heating in bushing connections	Loose connections due to improper handling	Check up the connections and ensure proper handling. Ensure free end clamp provision
(b) Low oil level and oil sample not satisfactory in case of non-communicative bushings	Due to leakage and non-maintenance of breather	Oil level to be topped up and oil to be changed with new oil and attending to oil leakage by replacing gaskets and maintaining the breather
	Note: Condenser bushings is hermetically sealed and only tanδ value to be periodically verified	
5. **Main tank and top cover** Leakage of oil resulting in depletion of oil level	Worn out gaskets weld leaks, leakage through valves	Gaskets to be replaced, re-welded after removing oil. Replacing valves or arrested with a plug roped with asbestos rope
6. **Oil handling systems**	(a) Improper oil level	Oil to be topped up in conservator
	(b) Lack of proper atmospheric pressure in conservator	Breathing system to be checked
	(c) Conservator	Conservator to be cleaned for removing sludges, iron particles and condensation
	(d) Misleading indication of oil level, leading to unnoticed depletion of oil. Revealed by MOG alarm or by Buchholz relay action of floats	1. Periodical check of oil level using dip stick 2. Checking up of increase/decrease of oil during loading operations 3. Improper oil indication by indicators to be attended
	(e) Broken gauge glass leads to entry of moisture	Glass indicator to be replaced
	(f) Change of colour of silica gel in breather indicate saturation of moisture absorption	Gel should be dried or changed
	(g) Defects in MOG alarm control circuit	DC control circuit to be checked periodically

7. **Cooling systems**	(a) Leakage through radiators (b) Header valves not opened before transformer put into operation; leads to temperature rise (c) Fan and pump not starting at desired temperature	To be arrested by proper welding after draining the oil and closing header valves Check the header valves to be opened before putting the transformer into service DC control circuit to be checked periodically and ensure proper cut off/cut in
8 **Thermometer**	(a) Non-availability of oil in thermometer pockets (b) Thermometers not showing correct temperature	Ensure oil in pockets Should be calibrated and adjusted
9. **Air release plug and equaliser pipe systems** Trap of air could not be identified. Leads to development of air pressure causing non availability of oil in certain pockets and causes damages.	 (1) During oil sampling (2) Due to filtration process	 Periodical release of air through air release plug and Avoid blocking in the equviliser pipe system
10. **Buchholz relay action**	i) Top float action – Due to minor fault (ii) Bottom float action – Due to major fault	Enumerated gas has to be tested and nature of defects assessed
11. **Vent pipe**	(i) Break of bottom diaphragm due to normal ageing ascertained by noticing oil in viewer glass (ii) Break of top diaphragm due to normal ageing (iii) Diaphragm breakages due to failure of transformer and oil spurt out	To be replaced soon since it is a must for Buchholz relay action To be replaced immediately since there is direct entry of moisture and dust Transformer to be completely examined and health checked

12. **PRV (pressure relief valve)**	(i) Non operation of PRV (ii) Non operation of flag indication (iii) Defects in control circuit	Action of PRV to be checked and its function to be ensured periodically Trip DC circuit to be checked and action ensured
13. **OLTC** (a) OLTC oil sample not satisfactory	(1) Due to more operation (2) Heavy carbonisation due to worn out contacts	Oil to be replaced periodically Contacts examined and worn out contacts replaced
(b) Surge relay operation	Heavy arcing inside chamber, leading to failure	OLTC has to be completely examined and measures taken
(c) Mixing of oil with main tank	Due to leakage in separator arrangement for tap leads entry	A separate conservator kept at a lower level than main tank conservator reveals this. Leakage to be arrested inside
(d) Problems in the operation of OLTC	Due to operation and control mechanism	To be rectified as per instructions manual of supplier

4

Operation of Power Transformers

This section describes the various protection schemes employed for the operation of power transformers. It covers the specific usage of auto transformers (single winding transformers) in the receiving stations associated with grid operation. The intricacies of the operation of transformers in a substation such as loading operation, tap operation and parallel operation are covered in detail.

4.1 Protection Schemes for Power Transformers and Their Operation

4.1.1 Introduction

The power transformer is the main equipment of the substation. It performs the important work of transforming power from one voltage level to another. The failure of a power transformer will lead to unnecessary interruption of power and revenue loss, as the existing loads will have to be fed by the capacity of available healthy transformers. Hence, it is necessary to protect transformers in all the possible ways. The types of protections employed for power transformers are discussed here.

4.1.2 Types of Protection

The protections based on electrical quantities are: (a) differential protection and (b) overcurrent protection. As far as radial substations are concerned, voltage reference is not used for any protection. Hence, potential transformers are used for metering only, whereas current transformers are used for metering and protection.

The other types of protections employed are:

1. Buchholz protection for winding and surge relay protection for OLTC
2. Protection from over temperature
3. Protection from excess liquid pressure
4. Protection from high voltage high frequency surges

4.1.3 Differential Protection

The differential protection employed is a current differential. This is fast and has complete coverage for the winding and terminal faults on the power transformer. The fundamental and general aspects are covered here.

General principle

The equipment to be covered under differential protection is shown with a CT on the incoming side and a CT on the outgoing side. The CT secondaries are interconnected through pilot wires and a relay is placed in the middle. Thus, the jurisdiction for the zone of protection is the area between the two CTs. When normal current, say 500 A, is flowing through the circuit, the secondary connections are made such that the same secondary current flows through the pilot wire circuit and the net current flowing through the differential relay coils is zero. (Refer to Fig. 4.1.)

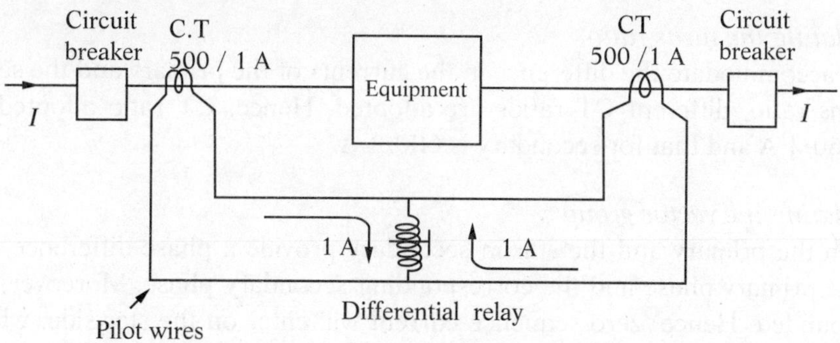

Fig. 4.1

But when the equipment is at fault, a heavy current will flow through the incoming side with no current on the outgoing side. Hence, the relay coil gets the actuating current and isolates the circuit by tripping the circuit breakers on either side of the equipment.

Application to power transformer

Coming to the application of transformers, the working conditions are different from the example shown above in that:

1. The current in the incoming side (primary) is different from the current on the outgoing side (secondary) based on voltage transformation ratio for maintaining different voltage level (*turns ratio*).

2. The connection of the primary winding is different from the secondary winding (*vector group*).
3. Heavy fault on the outgoing circuit that results in passing of huge current on the primary and secondary CTs differ in characteristics due to saturation, and the relay may operate (*through fault*).
4. Initial charging of the transformer results in heavy inrush current on the primary CT, with the current on secondary side being zero. In such conditions, the current balance will not be there (*inrush*).

Hence, arrangements must be made in the CT secondary connections employing a differential relay in order to accommodate the variation due to the turns ratio and vector group.

Moreover, inrush current and through-fault conditions are the phenomena outside the purview of failure of the transformer. Hence, these should be taken care of during the operation of the power transformer. The isolation of transformer should be done only for the fault on the transformer alone. They are achieved as given below.

Consider an example of 10 MVA 110/11 kV power transformer with 4_1DY11 connections and the current ratio of 52.5 A / 525 A.

Accommodating the turns ratio

In order to accommodate the difference in the currents of the primary and the secondary due to turns ratio, different CT ratios are adopted. Hence, CT ratio adopted for the primary is 60/1 A and that for secondary is 600/1 A.

Accommodating the vector group

The delta in the primary and the star in secondary provide a phase difference of +30° between the primary phase and the corresponding secondary phase. Moreover, the star point is grounded. Hence, zero sequence current will enter on the star side, whereas in delta side it will be circulated in the winding itself. Therefore, for an external through-fault, current balance will not be obtained as the zero sequence currents are not balanced in the CT outputs.

Hence, to accommodate this, the primary side CTs are connected in star where the transformer winding is in delta. The secondary side CTs are connected in delta where the winding is connected in star, with the star point earthed. Since Δ connection of CTs makes the pilot wire currents to be $\sqrt{3}$ times of secondary current, correction should be applied by employing auxiliary CTs of ratio $\sqrt{3}/1$ so as to balance with the CT current of the primary side.

Discriminating between through-fault and internal fault

This is achieved by biasing the relay through a restraining coil. During external fault conditions, the permitted small unbalance under normal condition will be magnified to a great extent, and so sensitive protection will suffer. Hence, to overcome this defect, a restraining coil usage is used. The restraining coil is placed in series in the middle of the pilot wire circuit, and the operating coil is placed across the two wires as shown in Fig. 4.2.

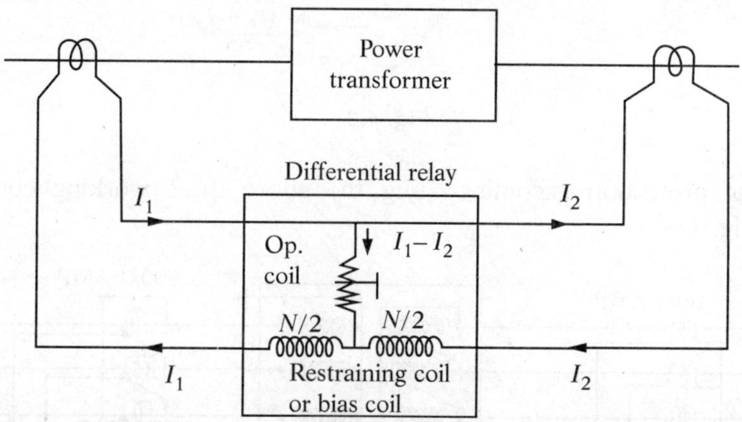

Fig. 4.2

By employing the restraining coil, the condition for the operation of the relay is that the torque due to operating winding should be greater than the torque due to restraining winding. The average restraining current is $(I_1 + I_2) \div 2$ in the whole restraining coil, whereas $(I_1 - I_2)$ is the operating current.

Hence, for external through faults, I_1 and I_2 increases and so the restraining force is greater than the operating force. The relay operation is thus prevented and selective discrimination is achieved. There will be tappings on the bias coil to adjust the setting of bias value (10% to 40%) and setting is also provided (50% to 100%) on the spring of restraining coil.

The operating characteristic of the relay is given in Fig. 4.3 for the operating force versus restraining force. This characteristic is a straight line as shown in Fig. 4.3. The operating characteristic of such a relay shows that the ratio of differential operating current to the average restraining current bears a fixed percentage. Hence, this is called *percentage differential relay* or *biased differential relay*.

The improved version of this is the self-adjusting bias type percentage differential relay. It has a high sensitivity at low value of restraining current, ensuring positive tripping in the case of light internal faults. Also, it has a lower sensitivity at higher value of restraining current, allowing CTs to depart from their true ratios under saturated conditions and prevent operation under through-fault conditions.

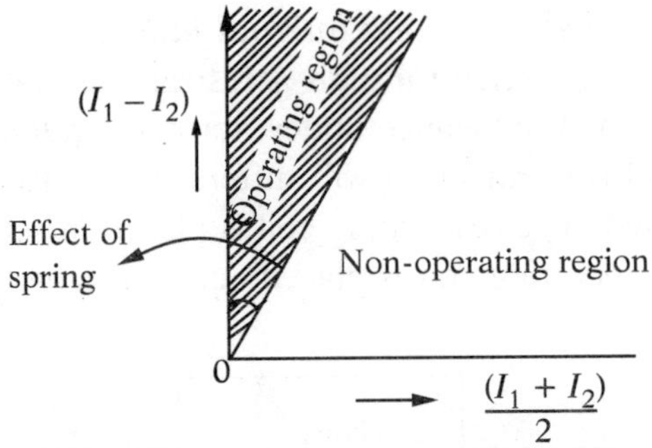

Fig. 4.3

The scheme of protection accommodating the above three working conditions are illustrated in Fig. 4.4.

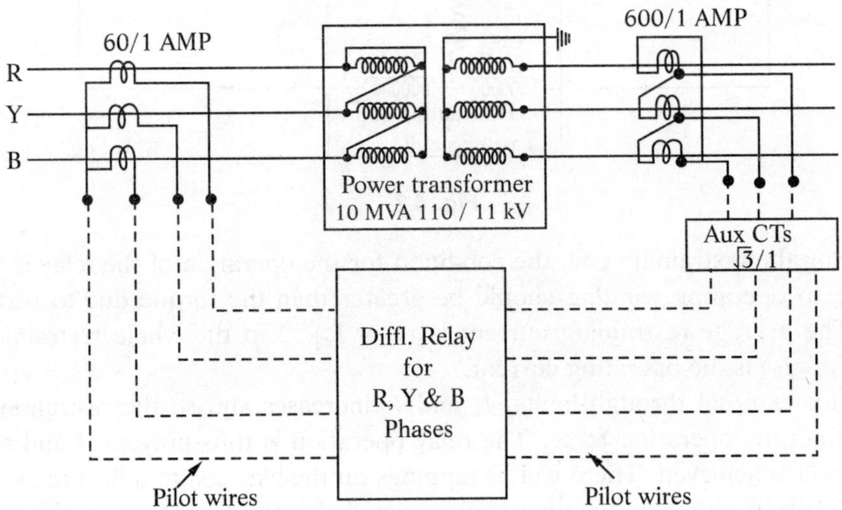

Fig. 4.4

The trip circuit of the differential relay is connected to the master relay for tripping all breakers in order to isolate the transformer from service.

Preventing operation of differential relay from inrush

The transformer winding being a highly inductive circuit, when switched on acts as a circuit without any impedance and the current has to be limited with the very low resistance of the circuit. Thus, heavy magnetising current flows for a very short duration and settles down to its normal value of 2 to 3% of full load current only after some time, as induced emf is built up in the winding.

This we call the magnetic inrush current of the transformer. Its value depends upon the point of the voltage wave at which it is switched on. If switched on when the wave is at zero point, a maximum inrush of 2 to 3 times the full load current develops. It also depends upon the polarity of the residual flux in the transformer winding. This heavy inrush current contains the fundamental AC magnetising current, its 2nd, 3rd 4th and 5th harmonic currents to an appreciable percentage, and a DC component.

The high harmonic current components are filtered, rectified and fed to the restraining coil while the fundamental component is fed to the operating coil. Thus the relay is made sensitive to fault currents but does not operate due to magnetic inrush. This is called harmonic restraint scheme and is shown in Fig. 4.5.

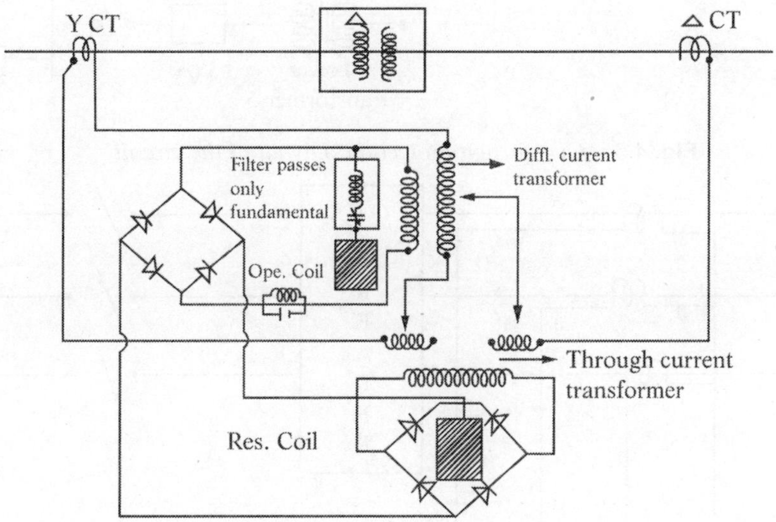

Fig. 4.5

4.1.4 Overcurrent Protection

The inverse current definite minimum time relays (non-directional) with high-speed instant trip is employed on the HV side of the power transformer as shown in Figs. 4.6 and 4.7.

Any fault on any of the transformers causes overcurrent to flow through the GC (Group Control) breaker, which is sensed by the CTs. The secondary of the CTs feed the O/L relays, which energises the trip circuit causing tripping of the GC breaker. This scheme also protects the transformer from normal overload on exceeding the permissible values by setting the relay suitably. For example, the total full load current on the HV side if all the three transformer banks are fully loaded comes to $3 \times 52.5 = 157.5$ A. The relay can be set to 110% with the adopted ratio of 150/1 A, with a time delay set to maintain O/L for a certain time.

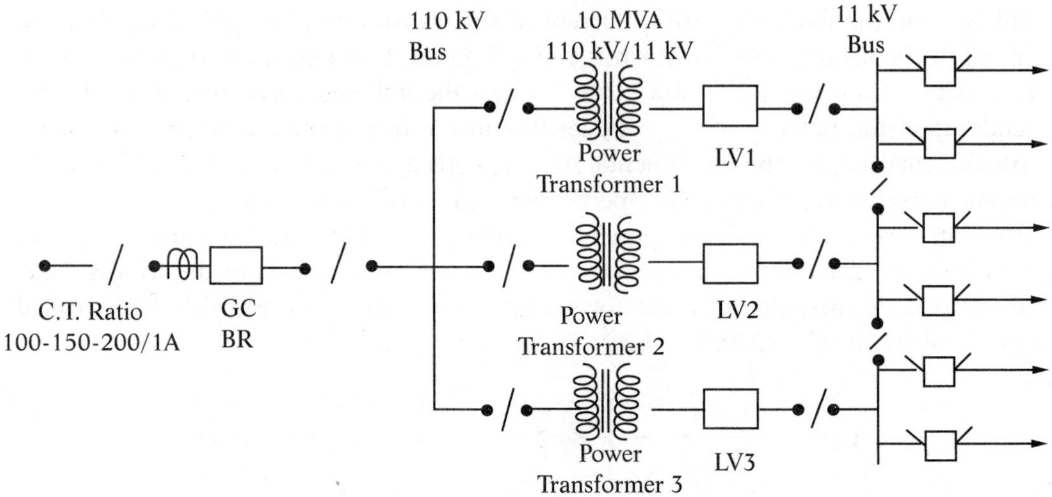

Fig. 4.6 *Overcurrent protection: Layout of the circuit*

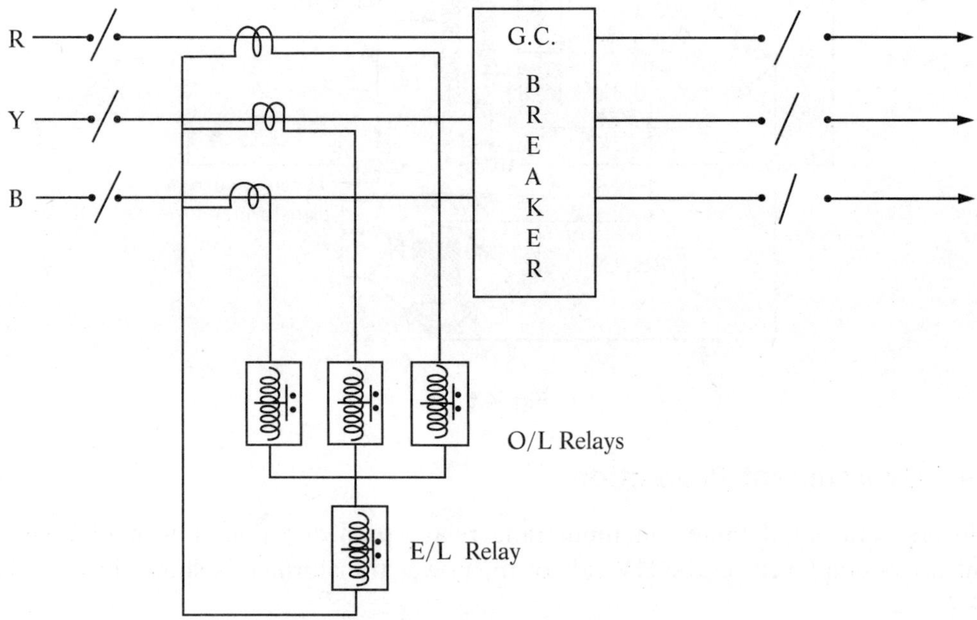

Fig. 4.7 *Scheme of protection for overcurrent*

Any fault on the feeder is taken care of by the IDMT non-directional relays associated with the circuit breaker operation employed for the feeders. Protection from the fault on 11 kV bus or backup protection for the feeder faults are taken care of by isolating the LV breakers of the transformer with a similar time-graded protection.

Thus, power transformers are prevented from feeding the fault, as faulty circuits are isolated well within the permitted time of the short-time rating of the transformer.

4.1.5 Buchholz Protection for Isolating Faulty Transformers

This is a gas actuated protective device. Any fault inside the transformer causes arcing and decomposes the oil. Turbulence is also created in case of heavy fault. The gas produced due to the decomposition of oil, burning of the insulating material, etc., and the turbulence or surge as it is called, are used for actuating a relay called the Buchholz relay. The placement of this relay and its construction details were explained in Section 3.1.9. Its protection features are described below.

(*i*) *Incipient fault on transformer*
In case of a small fault in the transformer, such as arcing in the connected leads or heating leading to burning of insulation, a decomposition of insulation and oil occurs and gas is developed slowly. This gas tries to enter the conservator where atmospheric pressure is maintained. While moving so, the gas gets collected and trapped in the top portion of Buchholz relay and pushes the oil downward. The oil will be pushed down due to such accumulation of gas, and when it goes down, the top float loses its buoyancy and falls down to the hinge, closing the contacts on the mercury bulb.

This fault is minor in nature and does not warrant immediate isolation of the transformer. Hence, the action of the top float is connected to the alarm circuit only. However, the faulty transformer identified is to be isolated manually for checking and rectifying the fault.

(*ii*) *Major fault on the transformer*
In case of a major fault, in addition to the enormous gas development there will be turbulence of oil and gas due to excessive pressure. The gas and oil rushes to the conservator. Here, the slow process of gas accumulation in Buchholz relay and then acting of the top and bottom floats is not justified. Hence, the bottom float is placed on the line of oil flow (line of flow of turbulence) from the main tank to the conservator, and a baffle is also attached to the float.

The turbulent oil and gas movement through the pipeline pushes the bottom float and makes the mercury on the bulb to come in contact, which bridges the terminals and energises the trip circuit, thereby isolating the transformer immediately. Because of this type of action, bottom float is also called *surge float relay*.

The bottom float contacts are connected to the master relay that trips all the breakers of the substation in order to isolate the faulty transformer and also prevent other healthy transformer banks from being overloaded.

The control circuit of Buchholz relay is given in Fig. 4.8 in a schematic form.

Since Buchholz relay is a gas-actuated relay, closed trip and alarm circuit will not be released when the faulty circuit is isolated unless the gas is released from the Buchholz relay chamber, unlike the differential or overload relays. Hence, a separate DC fuse, called *Buchholz relay fuse*, is provided to keep the trip circuit supply cut off on the defective transformer to enable restoration of supply to other healthy transformers. The gas analysis of the accumulated gas in the Buchholz relay will be done to ascertain the nature of fault, as discussed in Section 3.2.5.

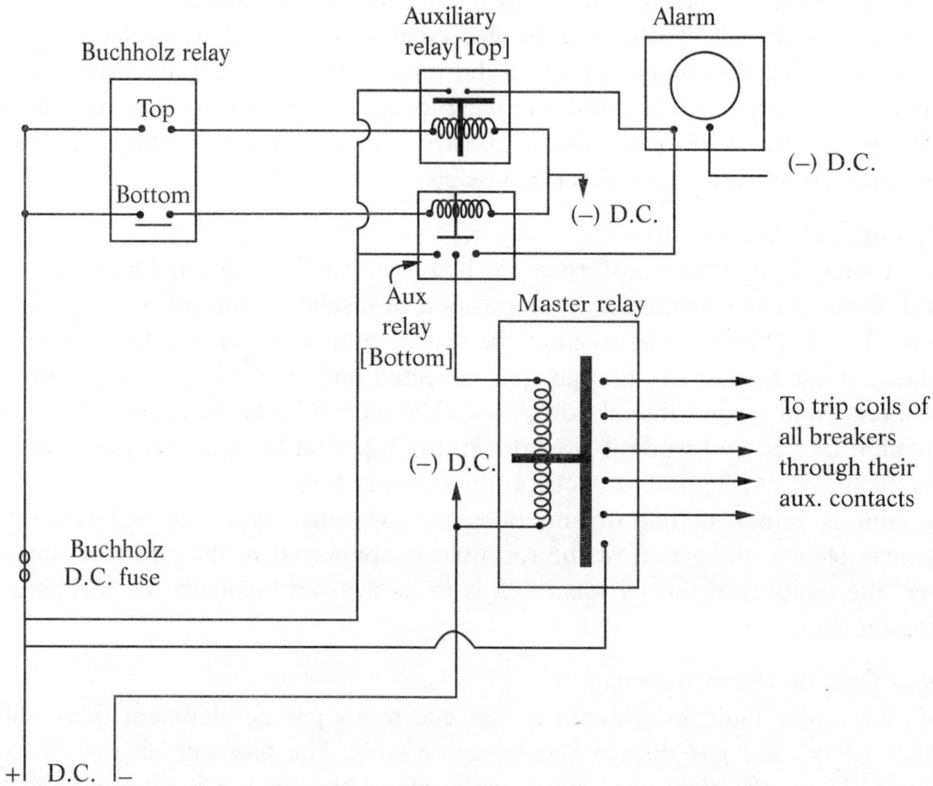

Fig. 4.8

Surge relay protection for OLTC

The diverter chambers are separate chambers housing the diverter switch contacts of the on-load tap changers immersed in oil. They are kept separated from the main tank oil to prevent the main tank oil from being polluted with carbonisation arising due to tap movement. Hence, problems may arise in this chamber due to failure of tap movements and turbulence is likely to be created, and so, this should also be protected. A surge float is provided in the surge relay (like bottom float of Buchholz relay), which is installed in the pipe connecting the OLTC chamber and its conservator.

The surge of oil and gas rushing to the conservator is made to act on this surge float, the contacts of which are connected to the trip circuit and annunciator. This arrangement is schematically shown in Fig. 4.9.

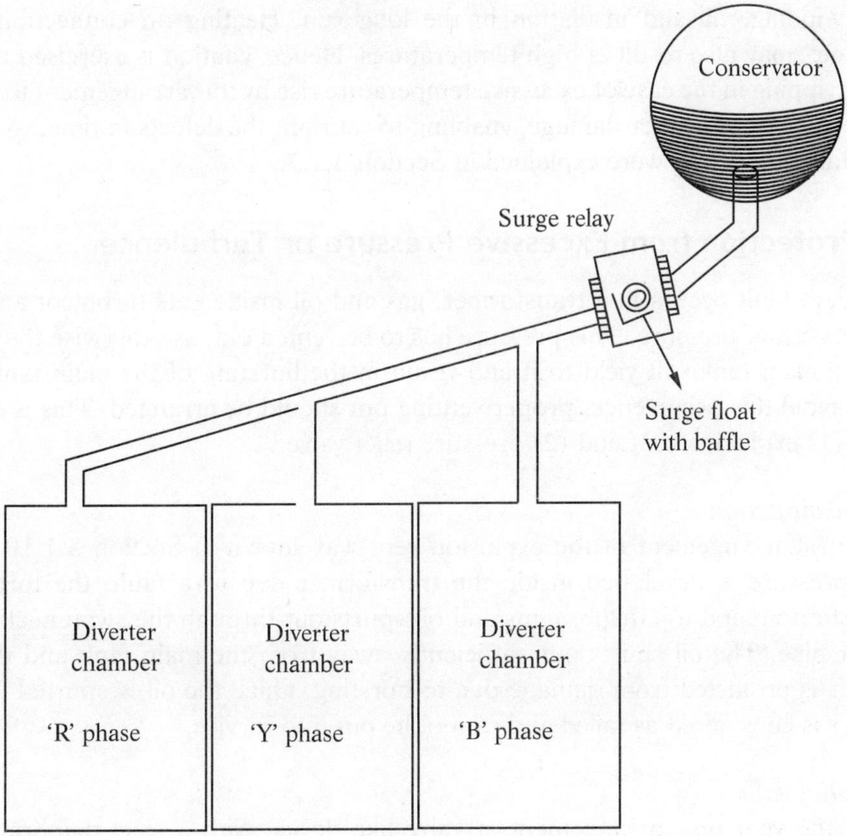

Fig. 4.9

4.1.6 Protection from Over-temperature (Thermal)

The power transformers are intended to operate on full load or at a small percentage of overload with a temperature rise of about 55° C for oil and 60° C for winding over the ambient temperature. The measurement of temperature of oil and winding are carried out using a thermometer with a capillary tube attached with the tube stem, which will be inserted in the oil pockets of oil and winding. The oil temperature is measured directly with the main tank oil, whereas the winding temperature is measured through a pocket of oil, with a resistance heating coil fed by winding temperature CT placed in the winding leads inside the transformer (*thermal image*).

The thermometers are provided with contacts, either directly or through mercury contacts, which can be set to operate at the required temperature. Usually, the oil temperature (max. 85° C) is connected to the alarm and the winding temperature (max. 95° C) is connected to the trip circuit.

An excessive winding temperature may result due to sustained overload, which may harm the winding, oil and insulation in the long run. Heating on connection leads, terminals, etc. may also result at high temperatures. Hence, caution is exercised through alarm and tripping in the case of excessive temperature rise by this arrangement to protect the transformer from further damage, enabling to set right the defects in time.

The installation details were explained in Section 3.1.7.

4.1.7 Protection from Excessive Pressure or Turbulence

When a heavy fault occurs in a transformer, gas and oil inside gets turbulent and gives rise to a very heavy pressure. This pressure has to be vented out, as otherwise the welded parts of the main tank will yield to it and result in the bursting of the main tank itself. Hence, to avoid this occurrence, proper venting out should be arranged. This is done in two ways: (1) explosion vent and (2) pressure relief valve.

Explosion vent action

The structural arrangement of the explosion vent was shown in Section 3.1.10. When excessive pressure is developed inside the transformer due to a fault, the turbulence breaks the bottom and top diaphragms and oil spurts out through the swan neck mouth of the vent pipe. The oil spurts out sufficiently away from the main tank and thus the transformer is protected from damage due to bursting. Once the oil is spurted out, the transformer is considered as failed and cannot be put into service.

Pressure relief valve

Instead of the vent pipe arrangement, a valve-like device with a steel diaphragm and spring, compact in size, is housed at a strategic location on the top of the main tank. This is also connected with a resetting device and annunciation contacts. The constructional details were described in Section 3.1.11.

Here, the excessive pressure operates on the steel diaphragm and moves the spring upward and opens the ports through which oil is spurted out. Such action also operates a mechanical arrangement that trips the station through the master relay. When excess pressure is released, the spring sets the PRV in its normal position. However, the fallen flag indicator should be manually reset to cut off the trip circuit to enable restoration of supply to other healthy transformers.

4.1.8 Protection from High Voltage, High Frequency Surges and Lightning

It is also necessary to protect the transformer from high voltage and high frequency surges in addition to protection from excessive currents arising due to fault.

(i) The transformer is protected from the direct stroke of lightning by erecting an *earth mast* on a structure near the transformer. The protective zone is a 45° cone with its apex as the vertex of the cone and a maximum radius equal to height of the earth mast from the circular base of the cone at the ground.

The transformer with its accessories should be within this protected zone. The earth mast should be properly grounded with very low resistance earth systems. This is shown in Fig. 4.10.

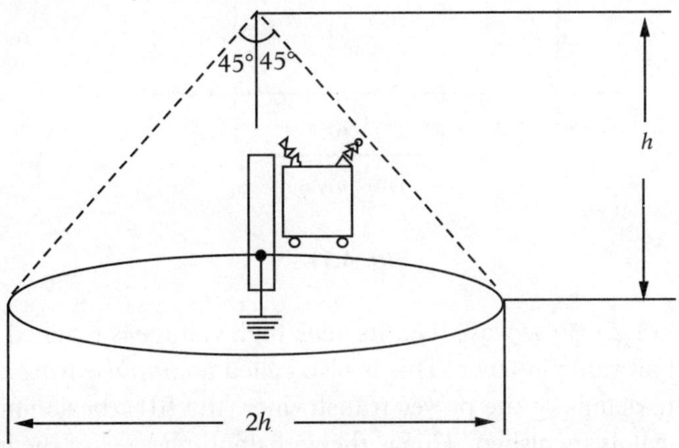

Fig. 4.10

(ii) The transformer is to be protected from high frequency and high voltage surges travelling on the line. These surges will arise due to

(a) External causes: Charges developed on the line due to static clouds over and above the line, which converts into high frequency, high voltage surges when a cloud discharges on another cloud or due to lightning striking near a line, causing a surge.

(b) Internal causes: Switching surges due to charging of an open ended line or cutting off of a heavily loaded inductive circuit.

These surges travel on the line in any direction as a travelling wave; the end of line in any direction meets in the transformer only. The line and bus insulations are heavy compared to the transformer winding insulation. So the winding of the power transformer is the easy target for this wave to smash and reach the ground.

Hence, the transformer is to be protected by channelising a proper path for this wave to reach the ground. Such a wave with a steep front is shown in Fig. 4.11.

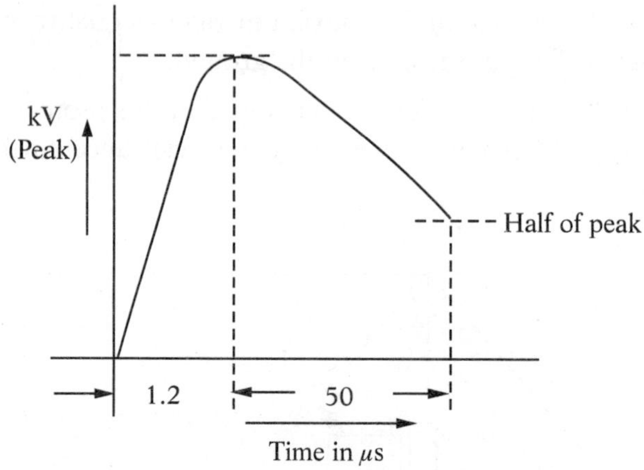

Fig. 4.11

The wave shown is a $1.2 \times 50\,\mu$s wave, i.e., its peak high voltage is reached at $1.2\,\mu$s, which settles down to its half value in $50\,\mu$s. This is also called an *impulse wave*.

In the name plate details of the power transformer, the BIL (basic impulse insulation level) value in kV peak is furnished. This is the withstanding level of the impulse wave by the transformer (for example, in 110 kV transformer; BIL: 950 kV peak of $1.2 \times 50\,\mu$s). Hence, surge waves coming on the line exceeding this level is to be diverted to ground and should not to be allowed to reach the transformer. This is accomplished by employing a *surge diverter* (erroneously called as lightning arrestor) on both HV side and LV side of the power transformer.

The surge diverter to be installed on either side of the power transformer should be well below the withstanding BIL value of the transformer. All the impulses coming on the line above this level will be diverted to ground by the surge diverters without allowing them to reach the power transformer. But, for the working voltage and frequency of the power supply, this will act as insulation. This is achieved by employing surge diverters with thyrite discs and gap units inside them. The surge diverters are to be installed within a stipulated electrical distance from the transformer bushing as shown in Fig. 4.12. Backup protection for this is also provided by rod gaps on the bushings of the transformer, which are also coordinated on the lines emanating from the substation on the terminal tower and on the three approach towers. The location of surge diverters with maximum electrical distances is shown in the figure.

The rod gap on transformer bushing for the 230 kV side is 1145 mm (45"), while for the 110 kV side it is 635 mm (25") and for the 66 kV side it is 395 mm (15.5"). No rod gaps are used on the 33 kV side and below on transformer bushing and on the lines since the clearance will be inadequate, making it prone to squirrel faults. The rod gap is shown in Fig. 4.13.

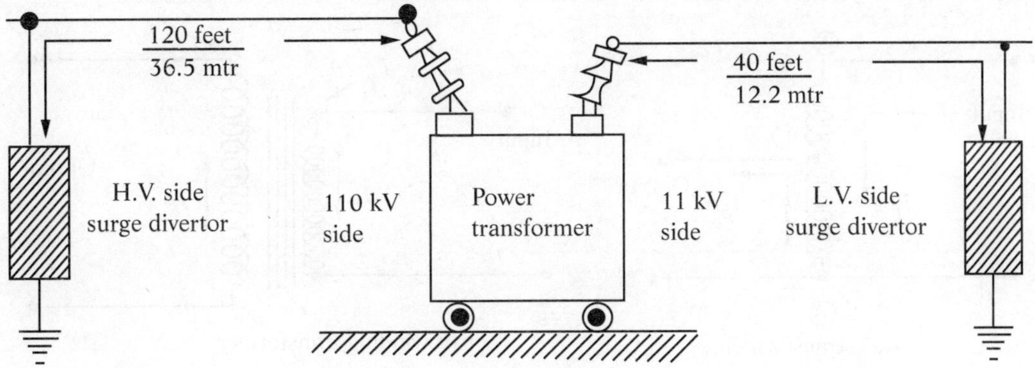

Fig. 4.12

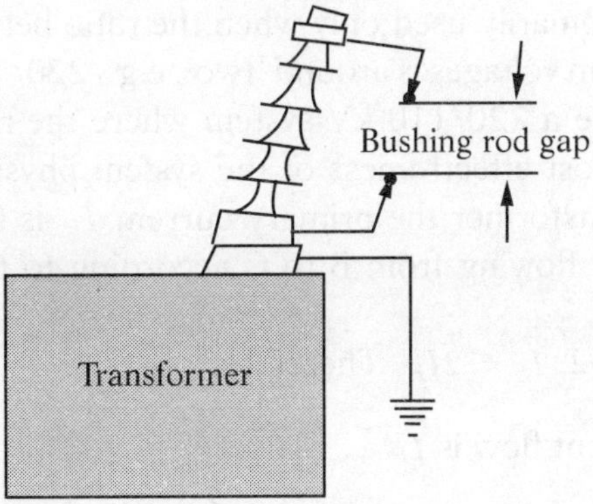

Fig. 4.13

4.2 Auto Transformer and Its Usage in Power Systems

4.2.1 Introduction

So far we have been discussing only the conventional two-winding transformers (i.e., with a primary and secondary winding. The functional requirement of the conventional two-winding transformers can be achieved with only one winding, without a secondary winding. Such a transformer is called auto transformer, based on the use of auto or variac principle. Its usage is resorted to on account of economical consideration only and not by virtue of any technical qualities.

4.2.2 Principle and Economic Considerations

The auto principle of an auto transformer is shown in Fig. 4.14 in a single phase representation along with a conventional two-winding transformer.

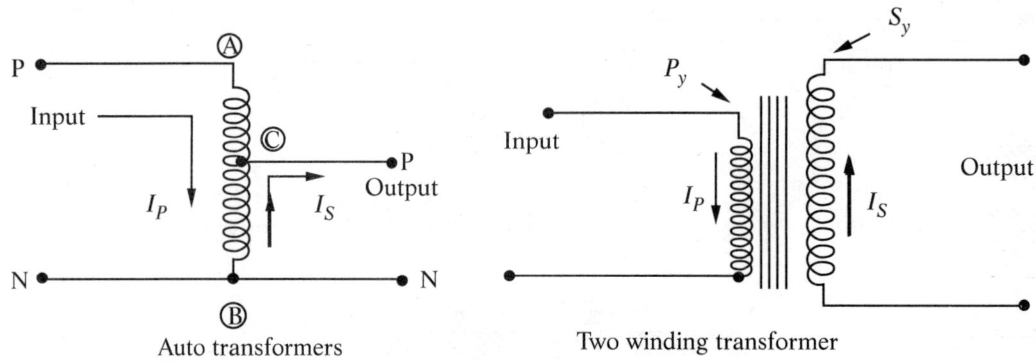

Fig. 4.14

Auto transformers are customarily used only when the ratio between the primary system voltage and secondary system voltages is around 'two', e.g., 220/110 kV stations, 400/220 kV stations, etc. Let us take a 220/110 kV system where the ratio of transformation is exactly 2 and analyse the cost effectiveness of the system physically. From Fig, 4.14, it is seen that in the auto transformer the primary current I_P is flowing from A to B and the secondary current I_S is flowing from B to C according to the transformer principle discussed in Section 1.1.1.

Since the ratio is exactly 2, $I_S = 2I_P$. Therefore,

From A to C Current flow is I_P

From C to B Current flow is I_P - $2I_P = -I_P$.

So, in the only winding available, for half the portion of the winding I_P is flowing. For the other half portion, again, only I_P flows but with a change in direction. Hence, it is enough if the winding is designed for handling the primary current flow I_P only. Thus, with the absence of a secondary winding, maximum economy is achieved.

This economical consideration is defined mathematically below. It should be noted that the percentage saving of copper winding becomes low when the ratio of voltage transformation increases from 2. When the ratio is more than 2, the auto-winding portion between A and C can be designed for the primary current I_P, but the portion between C and B should be designed for a current of $I_S - I_P$.

If the ratio is 3, then $I_S = 3I_P$, and therefore the design for the winding between C and B should be for $3I_P$ - $I_P = 2I_P$, whereby both the winding size as well as the cost increases. Hence, the cost consideration for the winding portion between C and B is made by comparing it with the secondary winding of a conventional two-winding transformer.

P_1 = Power supplied by the secondary winding of a two-winding transformer = $V_S I_S$.

P_2 = Power supplied by the winding between C and B of an auto transformer

$= V_S(I_S - I_P)$.

If ratio of transformation is $a = V_P/V_S = I_S/I_P$, then $P_2 = V_S(I_S - I_S/a)$ since $I_P = I_S/a$.

Material usage ratio $= \alpha = (P_2/P_1)$

$$= V_S(I_S - I_P)/V_S I_S$$

$$= (V_S I_S - V_S I_P)/V_S I_S$$

$$= V_S I_S/V_S I_S - V_S I_P/V_S I_S$$

$$= 1 - 1/a \text{ (since } a = I_S/I_P).$$

$$\text{Savings} = 1 - \alpha.$$

For various voltage ratios, the material usage ratio and the savings achieved are tabulated in Table 4.1.

Table 4.1

Voltage ratio a	Material usage ratio $\alpha = (1 - 1/a)$	Percentage savings achieved (1-α)
2	0.5	50%
3	0.67	33%
4	0.75	25%
8	0.875	12.5%
10	0.90	10%

It may be seen that cost savings by the use of auto transformers over the two-winding transformers drastically reduces with the increase in transformation ratio.

The following points have to be noted regarding the auto transformer windings.

1. When auto transformer is used for an exact ratio of 2, it is most economical.
2. There exists a physical interconnections between the incoming and outgoing system of auto transformer unlike the two-winding transformers, where the incoming and outgoing systems are electrically separate and linked only magnetically.
3. Since it is a common winding for both primary and secondary side, only graded insulation is required since the winding starts from zero voltage.
4. Absence of one winding reduces the percentage impedance and hence its inherent ability to protect itself from the short circuit stresses is poor.

4.2.3 Usage of Autos in Power Systems

The extra high voltage (EHV) interconnected network, called grid in power systems, necessitates the involvement of two or three EHV levels, for example 400 kV, 220 kV and 110 kV. In this operational system, the reduction of voltage levels and power flow is achieved using auto transformers as the transformation ratio is low. This is shown in Fig. 4.15.

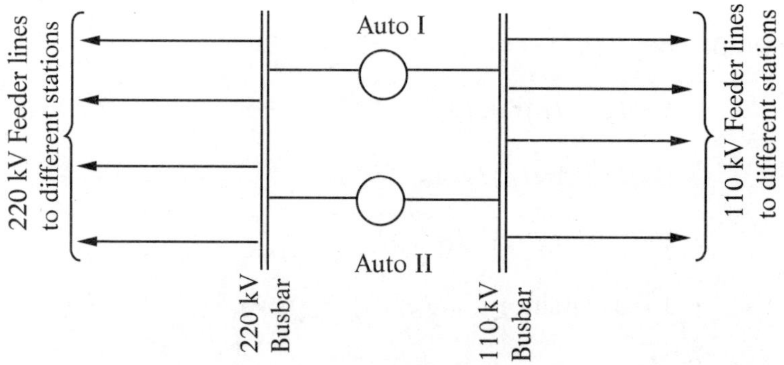

Fig. 4.15

Thus, autos can be considered as a link between the two different voltage levels of a grid. The power systems operation necessitates the provision of a neutral and earthing on the secondary side voltage level. It may be seen in the diagram that the 110 kV lines emanating from the 110 kV bus of the autos is connected to different station buses, with the autos providing the interlink.

Hence, all the autos provided in various stations should be of same vector group configuration to enable this tie. Thus, all the phase windings of an auto will be customarily kept connected in star with its neutral brought out on the 110 kV side, and hence uniformly Y0 vector group is provided on all autos. This configuration is schematically shown in Fig. 4.16.

4.2.4 Use of Tertiary Windings in Autos

Another requirement of the power system is to provide a delta winding at each transformation level for suppression of harmonics primarily, and for other operational considerations as discussed in Section 1.4.3. In conventional two-winding transformers, it is customary to have the primary winding connected in delta and the secondary winding in star, whereas in auto transformers there is only one winding kept connected in star. The absence of the delta winding connection, therefore, does not fulfil the operational consideration of a power system. Hence, it is customary to provide a separate winding on the same core limb of each phase, called 'tertiary winding'. These phase windings are kept connected in delta and its leads are brought out to a separate set of bushings. The delta connection may be either Yd_{11} or Yd_{1}.

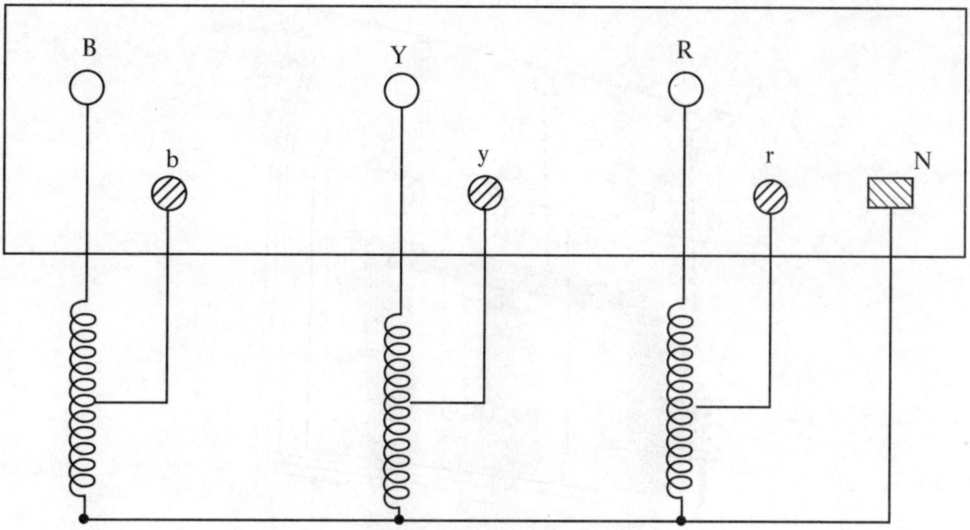

Fig. 4.16

Inasmuch as a separate winding has been provided as tertiary winding, this winding will also be designed to feed some loads as well. The capacity of this tertiary winding will usually be rated at about 30% of the rating of the main winding. For example, if the main winding of an auto is rated for 100 MVA, the tertiary winding will be rated for 30 MVA. The use of this tertiary winding is generally limited to capacitive loads for VAR compensation purposes only, as otherwise the loads catered to by the auto on 110 kV side will be reduced. This is discussed in Section 4.3.3.

When such external use is not resorted to, the tertiary windings will be kept unutilised but will continue to cater to harmonic suppression. It is also the practice in some utilities to keep one phase bushing of the tertiary winding connected to earth if the winding is not used externally. This is a protective measure for the winding, by maintaining the voltages with respect to ground. A schematic diagram (Fig. 4.17) shows the main and tertiary winding of autos and their interconnections.

4.3 Operation of Transformers in Substation

4.3.1 Introduction

Power utilities are the major bulk users of power and auto transformers for handling heavy power flow for supply to their customers. A discussion on the various operational problems encountered is important for engineers employed by them. The following discussion outlines the various operational features of transformers that an operational engineer needs to be conversant with to keep the system functioning normally.

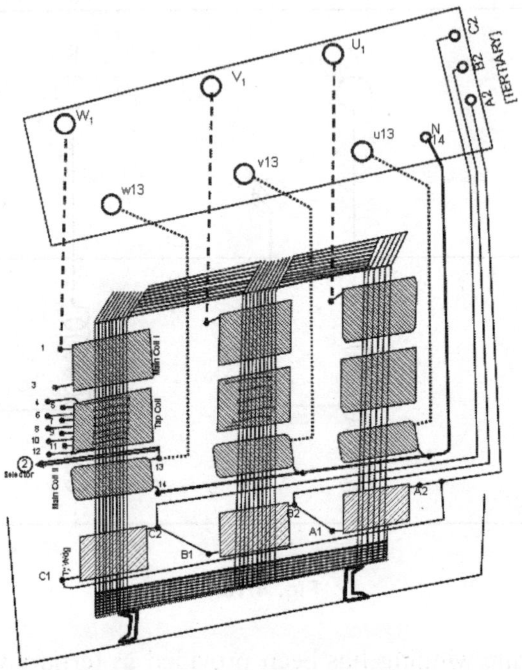

Fig. 4.17

4.3.2 Nameplate Details

The details given in the nameplate of a transformer guarantees and emphasises the operational possibilities. The nameplate details of a typical distribution small capacity transformer, a higher capacity power transformer used in sub-station and of an auto transformer used in grid network are given in Table 4.2. These details will be useful for the discussion on the operational aspects that follow. These details, provided by the supplier, have to be verified by the user by conducting pre-commissioning tests narrated in Section 2.2.0.

4.3.3 Loading Operation

The capacity of the transformer given in MVA in the nameplate details is the maximum apparent power that can be delivered by the transformer to the load side, as drawn from the source side, neglecting the losses. In case of power transformers, the power feeding capability will be defined only by the apparent power in MVA/ampere loading and not by the active power in megawatts since it is dictated by the power factor. In normal loading operation of the transformer, it can be loaded up to a maximum current defined in the nameplate. Any increase in loading beyond this limit is termed as overloading.

Table 4.2 *Nameplate details*

Sl No.	Details	Distribution transformer	Power transformer	Auto transformer
1	Capacity	100 kVA	10 MVA	100 MVA; Ty-30 MVA
2	Voltage ratio	11 kV/400-230 V	110 kV / 11kV	220 kV / 110 kV-11 kV(Ty)
3	Current ratio	5.25 A / 133 A	52.5 A / 525 A	251 A / 525 A (Main) 1575 (Ty)
4	Cooling system	ON	ON / ONAF 5MVA / 10 MVA	ON ONAF OFAF M 50 MVA 70 MVA 1000 MVA Ty 15 MVA 21 MVA 30 MVA
5	Guaranteed temp. rise	50° C	50° C	50° C
6	Tappings	5 Numbers (+5% to –5%)	17 Numbers (+5% to –15%)	17 Numbers (+10% to –10%)
7	% Impedance or % impedance volts	5%	9% at normal	Tap1 Normal Tap17 HV/LV: 15.072% 11.541% 9.723% HV/Ty: 63.578% 55.245% 48.352%
8	Vector group	4_1 DY 11	4_1 DY 11	HV/LV : Y0; HV/Ty : Y d 1
9	Power frequency withstand voltage	HV: 28 kV RmS	HV: 175 kV RmS LV: 70 kV RmS	HV: 395 kV RmS; LV: 230 kV RmS; Ty: 70 kV RmS
10	BIL (Basic Impulse Insulation Level)	HV: 75 kV Peak	HV: 900 kV Peak LV: 170 kV PEAK	HV: 950 kV Peak LV: 550 kV Peak Ty: 170 kV Peak
11	Quantity of oil	800 litres	11900 litres	44700 litres
12	Total weight	2.5 tonnes	39.30 tonnes	135.790 tonnes

Though the maximum MVA/current in the transformer is designed based on conductor sizes, the cooling system design is also another factor that influences loading. The nameplate details also give the various loading limits of the transformer depending upon the cooling systems (ON/ONAF/OFAF).

The rating of the transformer is for continuous rating and the guaranteed temperature rise specified is the maximum temperature rise over the ambient temperature when the transformer is loaded continuously for any number of hours. For example, if the ambient temperature is 35° C for a guaranteed temperature rise of 50° C, the maximum temperature that the oil will rise to is 85° C, and it will remain constant at full rated continuous loading of the transformer.

Overloading capabilities

In general, transformers should not be overloaded above the rated current specification. But in operation, it may be required to increase the loading more than the limit due to exigencies.

In practice, overloading is resorted to for a duration between half an hour to one hour. But then, the temperature rise is to be monitored and not allowed to go beyond the permissible oil temperature limit. Also, current should not be allowed to exceeded 10% of the rated current. The duration of overloading and consequent temperature rise depends upon the pre-overloading conditions of the transformer. If the transformer had been kept underloaded before overloading, the overloading duration will be more. But if it was kept at full load, the duration of overloading will be less due to the rapid rise in temperature.

In this case, if special cooling arrangement is resorted to, the possibility for sustaining the overloading for a longer duration can be thought of with oil temperature being maintained within the limit.

The operational engineer should decide the overloading possibilities based on the above conditions.

Overloading capabilities of power and auto transformer

The current rating of the transformer in its primary winding varies from tap No. 1 to tap 17 (in case of 17 taps). These variations are given below.

(i) For 10 MVA 110/11 kV power transformer

	Tap 1	Normal	Tap 17
Primary	50 A	52.5 A	61.7 A
Secondary	525 A	525 A	525 A

(ii) For 100 MVA 220/11 kV Auto transformer

	Tap 1	Normal	Tap 17
Primary	228 A	251 A	269 A
No separate secondary			

Hence, in case of a two-winding power transformer, though primary winding is designed to carry a maximum current of 61.7 A, the secondary winding is designed to carry 525 A only and thus the normal tap current of 52.5 A /525 A only should be considered as the full load rated current for overloading purposes. But in auto transformer, there is no separate secondary winding to impose such restrictions. Therefore, the maximum current in tap number 17 of 269 A itself can be taken as the full load rated current and the transformer loaded accordingly. The overloading can be considered accordingly with temperature limitations.

Loading of tertiary windings

Normally, in power system operation, the tertiary winding will not be loaded with MW (active power) since the system loading on the 110 kV side of the 220/110 kV station will be reduced. Hence, the tertiary winding will be loaded with var compensating devices, capacitors, static var compensators or synchronous condensers. Such MW loading in tertiary will reduce the active loading of the system through autos.

This is illustrated in the example below.

Auto transformer 220/110 kV : Capacity = 100 MVA

Tertiary windings 220/110 kV : Capacity = 30 MVA

Case I:

Loading of the autos in MW = 81 MW at 0.9 PF

Loading of tertiary = Nil

Apparent power loading of auto = 90 MVA

The power triangle is shown in Fig. 4.18.

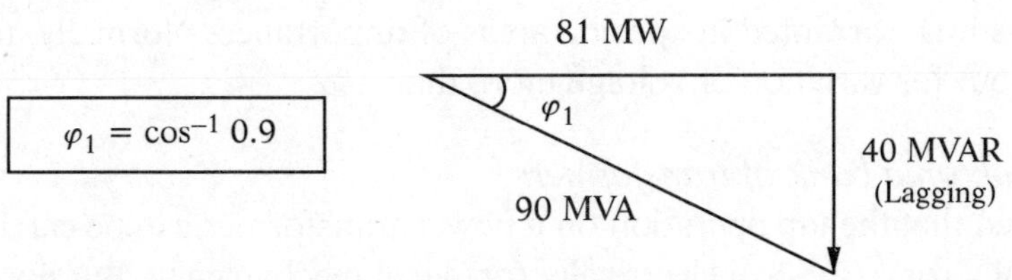

Fig. 4.18

Case II:

Loading of the auto in active power remains as 81 MW.

Loading of tertiary = 10 MVAR capacitors

The power triangle is shown in Fig. 4.19. By injecting 10 MVAR through the tertiary winding, the net MVAR loading is reduced to 30 MVAR and hence, for a sustained active power of 81 MW, the autoloading is reduced to 86 MVA.

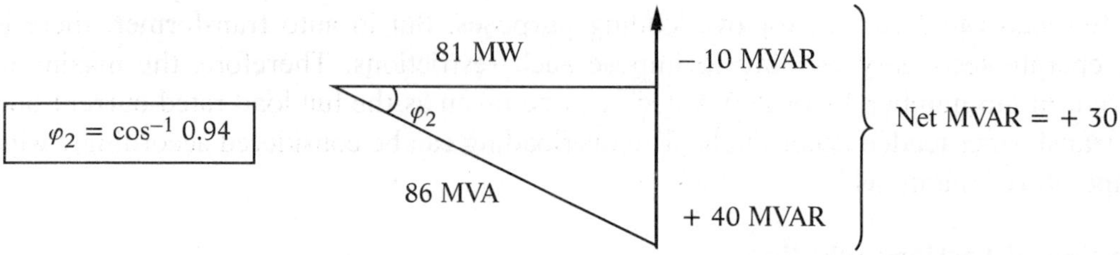

Fig. 4.19

4.3.4 Tap Operation

Although the details of tappings and tap operations in transformer were discussed in Section 1.5, some operational procedures are described here. The tap numbers with voltages were furnished in Table 1.5 for a 10 MVA 110/11 kV power transformer and for a 100 MVA 230/110 kV auto transformer.

(*a*) *Judicial selection of taps*

It may be seen in the details of taps that the secondary voltage should be kept constant for the variation of voltages at the incoming primary side. Hence, if the secondary voltage at the transformer decreases/increases, the quantum of this decrease/increase will decide the tap to which changing should be done to maintain the secondary voltage at the required level.

Change of tap should not be resorted to for making up *tail end voltage variations*.

The tap changing operation need not be resorted to frequently for very small variations of voltage unless it is warranted in specific areas of importance. Normally, tap operation may be carried out for variation of voltage more than 3%.

(*b*) *Tap operation on a bank of transformers*

It may be recalled that the tap operation on a power transformer can be carried out by (i) remote electrically (or), (ii) local electrically (or) local mechanically. But if a bank of 2 or 3 transformers are operated in parallel, it is necessary that all transformers be operated in the same tap to avoid flow of circulating current due to difference in voltage. Hence, it is customary to have Master/Follower/Independent control setting switches in each transformer panel.

One of the transformers have to be set at *Master* and other transformers at *Follower*. The tap operation should be done only on the master transformer and other transformers will follow the setting automatically. Any transformer can be selected for master control. In such operation, if other transformers do not follow, an out-of-step indication is provided to take corrective measures.

If the master/follower control is not available, the tap operations can be done independently as follows:

(i) The tap operation is very fast and hence keeping one tap variation between the transformers for that short duration while changing tap in the bank of transformers is generally allowed.

(ii) Hence it is not necessary to press the tap switch simultaneously on all the transformers but can be done 'one by one' by keeping only 'one tap difference' during tap changing.

(c) *Use of taps for control of momentary loading*

While the transformers are under full load and trying to take the overload, reducing one or two taps in this condition will reduce the loading to a certain extent and peak loads can be managed in case of night peak only. This should not be resorted to in the morning peak, since this is dictated by power factor.

4.3.5 Parallel Operation of Transformers

In the operation of power system, three types of operations are resorted to.

(a) *Radial operation* The power supply is fed through one source on the primary side of the transformer and its secondary independently feeds the tail end. This is followed in distribution systems.

(b) *Parallel operation* A bank of transformer is fed through one source but the secondaries will be paralleled and will feed the system radially. This is practised in radial sub-stations.

(c) *Synchronised operation in interconnected grid network* The different voltage levels of transmission network (for example 400/220 kV or 220/110 kV) are synchronised through a bank of auto transformers; the network is called *grid network.*

No description is necessary for radial operation but for the other two, special significance is to be assigned by choosing proper parameters consistent with paralleling or synchronising.

Parallel operation of transformers

The parallel operation of transformers is carried out by linking the corresponding secondary phase of all the transformers. Hence, the following parameters of the transformers should be same.

1. Identity of R, Y and B phase
2. Voltage ratios with identical tappings
3. Percentage impedance
4. Vector groups

It is not necessary that the capacity of all transformers be the same. In practice, a capacity ratio of 1:3 can be adopted for the transformers in the bank. In sub-stations, the phase identity of line conductors is traced from the source and is available at the bus of the station. Only voltage ratios with identical tappings are installed. Hence, detailed analysis are made for the last two parameters.

Parallel operation with equal percentage impedance

The percentage (or per-unit) impedance is the guiding factor and not the ohmic value of impedance of transformers, as discussed in Section 1.1.13. If the transformers are of equal percentage impedance, the loading will be shared according to the capacity of the transformers. If a load of 26 MVA is catered to, then 10 MVA Unit-I will deliver 10 MVA load and the 16 MVA Unit-II will deliver 16 MVA load (Fig. 4.20).

The load sharing can be ascertained by using the following formula given below.

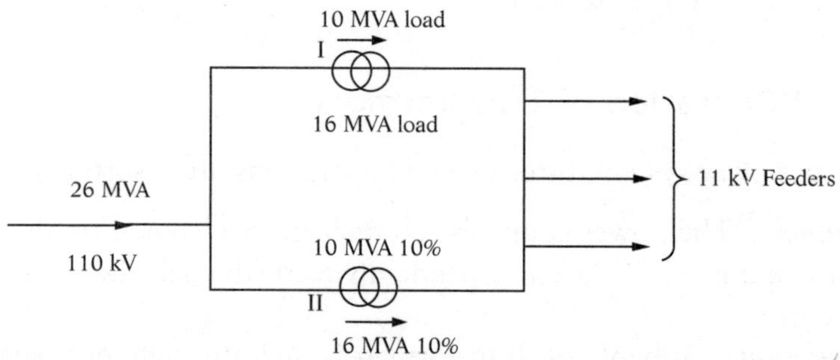

Fig. 4.20

Load sharing formula for parallel operation

Let S = Total supplied load in MVA

S_1 = Load shared by transformer I in MVA

S_2 = Load shared by transformer II in MVA

Z_1 = Percentage impedance of unit I of capacity 10 MVA

Z_2 = Percentage impedance of unit II of capacity 16 MVA

Since Z_1 and Z_2 are specified with the capacity of their unit bases, a common capacity base should be considered for calculations. Let us assume a 100 MVA base.

Let Z_{11} in the % impedance be converted to the new base of 100MVA. Then,

$$Z_{11} = Z_1 \times \frac{100}{10} = 10Z_1 = 10 \times \frac{10}{100} = 1 \text{ p.u.}$$

Similarly, $Z_{12} = Z_2 \times \dfrac{100}{16} = 6.25\, Z_2 = 6.25 \times \dfrac{10}{100} = 0.625 \text{ p.u.}$

Now, $S_1 = S \times \left[\frac{Z_{22}}{Z_{11} + Z_{12}} \right]$

$$S_2 = S \times \left[\frac{Z_{11}}{Z_{11} + Z_{12}} \right].$$

The total load of 26 MVA given in the example will be divided as follows:

$$S_1 = 26 \times \frac{0.625}{1.625} = 10\text{MVA}$$

$$S_2 = 26 \times \frac{1}{1.625} = 16\text{MVA}$$

Note: The above formula is given below for 3 transformers of a bank, which will be significant and useful in practical application sw.

$$S_1 = S \times \left[\frac{Z_{22}Z_{33}}{Z_{11}Z_{22} + Z_{22}Z_{33} + Z_{33}Z_{11}} \right]$$

$$S_2 = S \times \left[\frac{Z_{33}Z_{11}}{Z_{11}Z_{22} + Z_{22}Z_{33} + Z_{33}Z_{11}} \right]$$

$$S_3 = S \times \left[\frac{Z_{11}Z_{22}}{Z_{11}Z_{22} + Z_{22}Z_{33} + Z_{33}Z_{11}} \right]$$

Parallel operation of transformers with unequal percentage impedance

Even though parallel operation of transformers has to be carried out with equal percentage impedance only, sometimes, situation will warrant paralleling of transformers with unequal impedances also (Fig. 4.21). In this situation, the load sharing will be different and it will not be according to their rated capacities. The load sharing is discussed below for the two transformer banks with unequal percentage impedances.

$$Z_{11} = Z_1 \times \frac{100}{10} = 10 \times \frac{9}{100} = 0.9 \text{ p.u.}$$

$$Z_{22} = Z_2 \times \frac{100}{16} = 6.25 \times \frac{10}{100} = 0.625 \text{ p.u.}$$

$$S_1 = 26 \times \frac{0.625}{1.525} = 10.7 \text{ MVA}$$

$$S_2 = 26 \times \frac{0.9}{1.525} = 15.3 \text{ MVA}$$

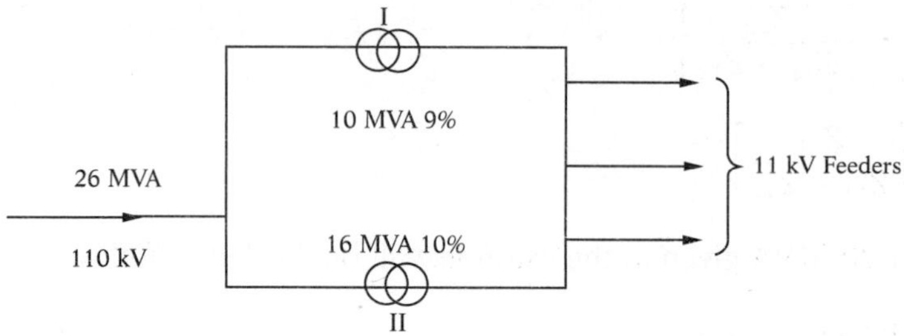

Fig. 4.21

We find that transformer 1 is overloaded while transformer 2 is under-loaded. Thus, the inference is that a low percentage impedance transformer will be overloaded while a high percentage impedance transformer will be under-loaded. The transformers cannot be loaded to their full capacity and loading should be limited such that the low percentage impedance transformer is loaded to its full capacity, and automatically, the higher percentage impedance transformer is under-loaded. In such paralleling condition, only the loading of lower percentage impedance transformer need be watched, keeping it as reference.

Circulating current concept

The different loadings are shown in Fig. 4.22. Let us assume that the rated current is flowing in the winding. Since percentage impedance is different, the voltages V_1 and V_2 will be different. In order to keep the voltage at the same level, a circulating current of 37 A flows in the secondary winding of both transformers. But this circulating current flows in the same direction of load current in Transformer 1, and so adds with the rated current resulting in net loading as 562 A, i.e., 525 + 37. It flows in the opposite direction in transformer No. 2 and results in reduced current with net loading as 903 A, i.e., equal to 940-37.

Inference on losses

1. Rated load current produce normal losses at full load.
2. The circulating current of 37 A in the secondary winding of transformer No. 1 and transformer No. 2 is to be considered as additional current flow.
3. This additional current of 37 A, irrespective of the direction, produces additional heat in the two secondary windings through resistances $RT1$ and $RT2$.
4. Hence, additional losses in 3 phases of winding is

$$\frac{3 \times 37^2 \times [RT1 + RT2]}{1000} \text{ kW.}$$

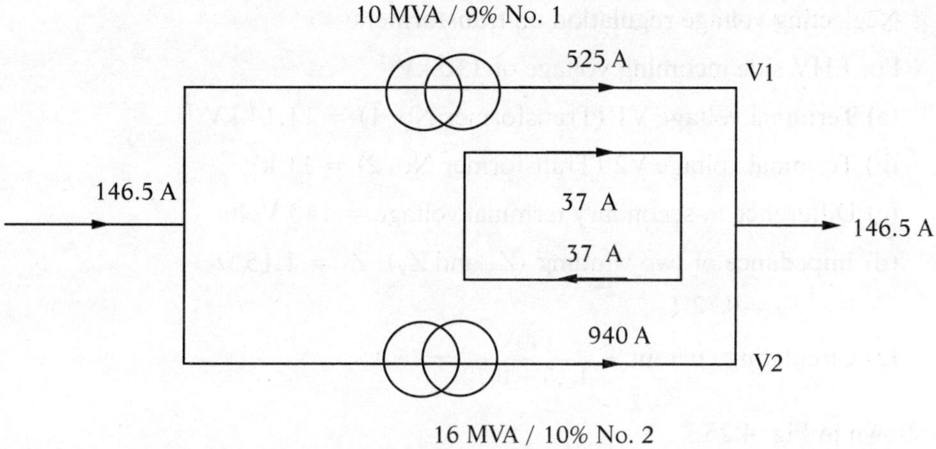

Fig. 4.22

Effect of operation of two transformers with equal impedances with different tappings

Consider two transformers with equal percentage impedance of 10% with details of tapping as follows:

1. Transformer No. 1
   10 MVA 110/11 kV: Tap +5% to –15% (17 taps). Each Tap = 1.25% (1375V) HV side tapping

2. Transformer No. 2
   16 MVA 110/11 kV : Tap +5% to –15% (17 taps). Each Tap = 1.25%(1375 V) HV side tapping

3. System conditions:
   (a) System voltage at EHV Side : 110 kV
   (b) 10 MVA transformer No.1 : Tap changed from tap No. 5 to 6
   (c) 16 MVA transformer No.2 : Tap kept at tap No. 5 (Existing) (Tap not changed)
   (d) Turns ratio for tap 5 = 10
   (e) Turns ratio for tap 6 = 9.875

4. Neglecting voltage regulation on transformers:

For EHV side incoming voltage of 110 kV

(a) Terminal voltage V1 (Transformer No. 1) = 11.14 kV

(b) Terminal voltage V2 (Transformer No. 2) = 11 kV

(c) Difference in secondary terminal voltage = 140 Volts

(d) Impedance of two winding (Z_1 and Z_2): Z_1 = 1.15 Ω

$Z_2 = 0.72\Omega$

(e) Circulating current $= \dfrac{140\text{V}}{1.15 + 0.72} = 75$ A

This is shown in Fig. 4.23.

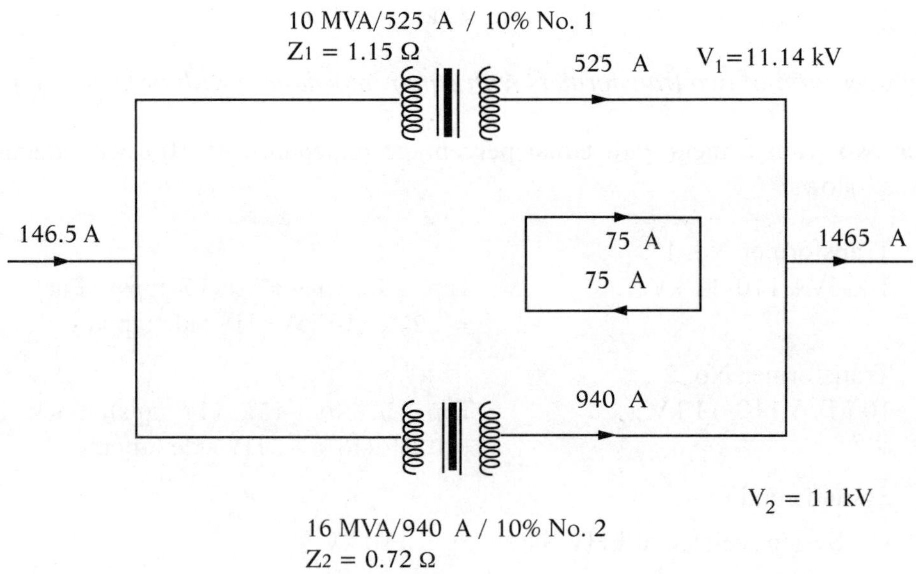

Fig. 4.23

Loss occurrence and other precautions

1. Just like percentage impedance difference, make a voltage difference between V_1 and V_2. Different tap settings make a voltage difference of 140V between V_1 and V_2.
2. Difference in voltage make a circulating current of 75 A.
3. This circulating current gets added to the full load current of transformer No.1 and opposes the direction in transformer No.2.
4. Net loading of transformer No.1 is 600 A and that of transformer No. 2 is 865 A.
5. This circulating current produces additional losses equal to

$$\frac{3 \times 75^2 \times [R_1 + R_2]}{1000} \text{ kW}$$

6. Master–Follower controller is essential.
7. If not available, tap changing may be done keeping one tap difference only at a time, and thus bringing to equal tap numbers.

Permanent parallel operation of transformers irrespective of load pattern

Consider two 10 MVA transformers of 110/11 kV with 10% impedance each (Fig. 4.24).

No load loss for each transformer	=	20 kW
Full load loss	=	300 kW
Resistance of winding/phase (Transferred to secondary)	=	0.33 Ω

Equation for total losses for different load factor:

No load losses + Full load losses × $(\text{L.F})^2$

Load Factor (L.F) for each transformer can be calculated.

Total load = 8 MVA [within one transformer capacity]

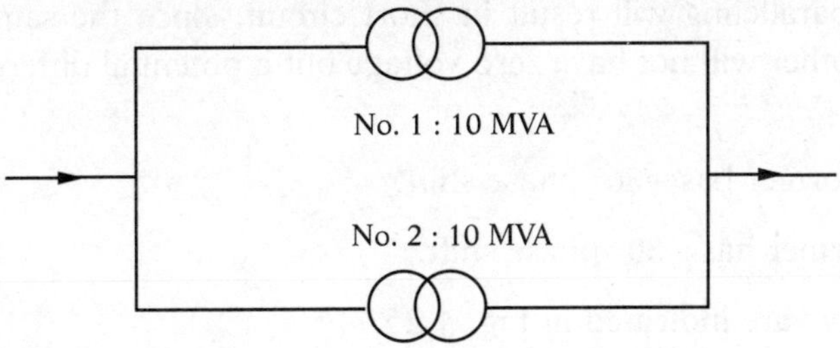

Fig. 4.24

Case 1

If transformer is operated in parallel,

L.F = 0.4 for each transformer.

Then,

$$\text{Loss} = [\ 20 \text{ kW} + 300 \text{ kW} \times (0.4)^2] \times 2 \text{ transformers}$$

$$= [20 \text{ kW} + 48 \text{ kW}] \times 2 = 136 \text{ kW}.$$

Case 2

Total load of 8 MVA is fed by Transformer No. 1 only.

Transformer No.2 is switched off.

L.F = 0.8

Then, Loss = [20 kW + 300 kW $\times(0.8)^2$] $\times$ 1 transformer

= [20 kW + 192 kW] $\times$1 = 212 kW.

Conclusion

(a) From the loss point of view, even though the load is within one transformers capacity, it can be fed by both transformers in parallel so that there is a loss saving of 212 kW – 136 kW = 76 kW.

(b) This will also provide heat run for transformers.

Importance of similarity of vector group

The similarity of vector group of transformers while paralleling is a must. During paralleling, the primary side is fed by the same source but the secondary side undergoes a phase shift because of the interconnections as discussed in Section 1.4.5.

Hence, it is necessary that all transformers to be paralleled should be in same vector group. Invariably, 4_1DY11 vector group is adopted for all transformers. If vector groups are different, then paralleling will result in short circuit, since the same phases of one transformer and another will not have zero voltage but a potential difference.

For example:

4_1DY11 transformer has +30° phase shift.

3_1DY1 transformer has –30° phase shift.

The secondary vectors are indicated in Fig. 4.25.

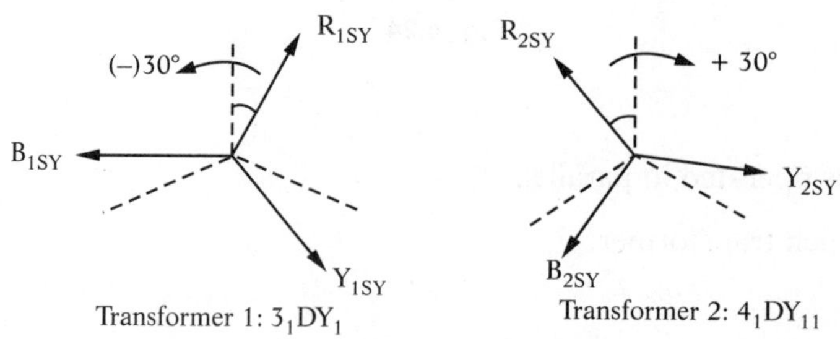

Fig. 4.25

Paralleling means bridging R1 phase with R2 phase, y1 phase with y2 phase and B1 phase with B2 phase. Combining the two, we find R1 phase has a phase shift of 60° with the

R2 phase vector. Thus, since it is an equilateral triangle, assuming these transformers are 110 kV / 11 kV units, then R1 and R2 (and also other similar phases) have a voltage difference of 11 kV÷√3 volts. Hence, paralleling will result in short circuit.

Therefore, in general, transformers with different vector groups cannot be paralleled. The 4_1DY11 and 3_1DY1 vector groups can be changed by interchanging the delta connection if the terminals are brought below the top cover, as already discussed in Section 1.4.6.

Changing of bus feeding arrangements to parallel 4_1DY$_{11}$ and 3_1DY1 transformer

As a special case, the other method of nullifying the effect of 4_1DY11 and 3_1DY1 can be resorted to by changing the phases of the bus connection on the primary side and the secondary side of one of the transformers as shown in Fig. 4.26. But 'caution' has to be exercised by re-changing the bus connection while replacing the 3_1DY1 transformer with the usual 4_1DY11 unit.

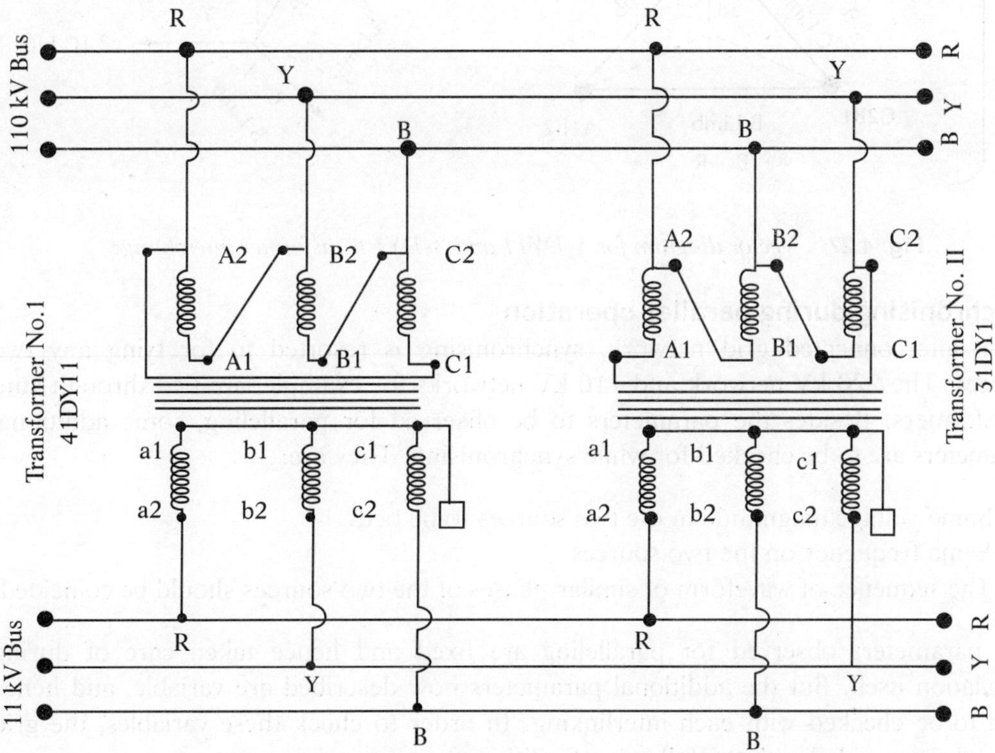

Fig. 4.26 *Bus interchange arrangement for parallel operation of 4_1DY11 and 3_1DY1 transformer*

The vector diagram for the above interchange is shown in Fig. 4.27.

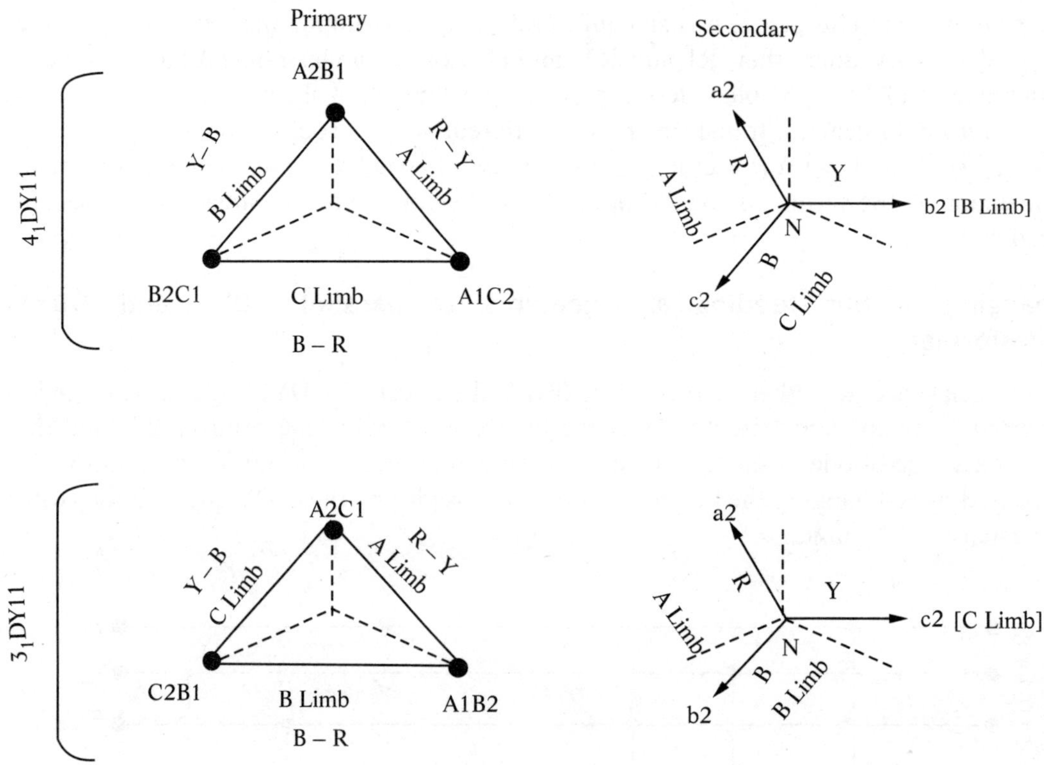

Fig. 4.27 *Vector diagram for 4_1DY11 and 3_1DY1 transformer interchange*

Synchronising during parallel operation

In an interconnected grid network, synchronising is resorted to for tying any two sources. The 220 kV network and 110 kV networks for example, are tied through auto transformers. Besides the parameters to be observed for paralleling, some additional parameters are to be checked for while synchronising. They are:

1. Same voltage magnitude in the two sources to be tied
2. Same frequency on the two sources
3. The sequence of waveform of similar phases of the two sources should be coincided.

The parameters observed for paralleling are fixed and hence taken care of during installation itself. But the additional parameters now described are variable, and hence need to be checked with each interlinking. In order to check these variables, the grid network is provided with the following facilities.

1	Two voltmeters	One for existing source and the other for the source to be tied
2	Two frequency meters	One for existing source and other for the source to be tied
3	One synchroscope	Connected to the two sources for verification of the coincidence of waveforms This will be indicated by the pointer pointing to 12 o'clock position. If this pointer is not stable at 12 o'clock position but moves in clockwise direction or anticlockwise direction, it means the waves are not coinciding.

In our discussion of auto transformers, the parallel operation of transformers discussed before holds good for a bank of auto transformers also. Hence, loading conditions, percentage impedance variations, etc., during parallel operation has to be observed.

But the special operational activities during the parallel operation of autos are:

1. If two autos, for example, are kept in parallel, the removal of one auto alone and cutting-in the same after works does not involve any special activity; they can be paralleled as it is without synchronising, since one auto is already kept tied between the 220 kV and 110 kV systems.
2. If all the two autos are cut off, then the 220 kV system and 110 kV system tie is broken and hence, while cutting in back, the first auto should be synchronised observing the synchronising formalities, but the second auto can simply be paralleled.

Equal impedance setting using taps

The impedance of transformer varies from tap to tap as seen from the name plate details. If different makes of transformers with different percentage tapping are to be paralleled in special circumstances, each transformer can be selected with a tap having equal percentage impedance. Then the transformers can be paralleled even though each transformer will be set at different taps in order to make equal percentage impedance setting.

In this case, ON load tap changer operation should not be resorted to and all transformers should be operated as 'fixed tap' units. The percentage impedance on various taps, if not available in nameplate details, has to be ascertained for all taps as described in Section 2.2.7.

4.3.6 Emergency Operation of Transformers

The in service transformers in a system may face some emergency situations also. The emergency considered may be as below:

1. The transformers should not feed any fault in the system network. The faulty

network should be isolated and other healthy systems continued to be fed. This is called *through fault.*

2. A transformer may become defective and in such a case, the defective transformer has to be isolated from the system.

Proper protective schemes are employed to meet the emergencies as described in earlier in Section 4.1.0. The operational aspects of the schemes are described below in brief.

Through-fault emergency

An overload and earth leakage protection is employed on the LV side and the HV side of the transformer bank, and thereby the faulty system is isolated.

Fault on transformer

The protective schemes are:

(A) The emergencies which are not that serious in nature are indicated by the following schemes through alarm indications only. They are:

<table>
<tr><td>(i)</td><td>Exceeding oil temperature level</td><td>Oil temperature alarm</td></tr>
<tr><td>(ii)</td><td>Low oil level</td><td>MOG alarm</td></tr>
<tr><td>(iii)</td><td>Incipient small fault on joints or unusual hotspot</td><td>Buchholz top float action alarm</td></tr>
</table>

For oil temperature alarm, it is enough if loads are cut off since it happens due to overload only. But for items (ii) and (iii) above, an emergency condition results. Hence the transformer will have to be cut off from service and the emergency attended to. For MOG alarm, the oil should be topped up to the required level and then the transformer put back into service. For item (iii), the transformer has to be completely inspected after carrying out tests on the gas collected in Buchholz relay as described in Section 3.2.6.

(B) The emergency arising due to major fault on the transformer, which is serious in nature, is indicated by the following protection schemes either independently or by all schemes.

<table>
<tr><td>(i)</td><td>Current differential relays</td><td rowspan="3">Electrical quantity based actuation</td></tr>
<tr><td>(ii)</td><td>Overload relays</td></tr>
<tr><td>(iii)</td><td>Over flux relays</td></tr>
<tr><td>(iv)</td><td>Buchholz bottom float</td><td rowspan="3">High pressure oil and gas surges based actuation</td></tr>
<tr><td>(v)</td><td>OLTC surge float</td></tr>
<tr><td>(vi)</td><td>Pressure relief valve</td></tr>
<tr><td>(vii)</td><td>Winding temperature relay</td><td>Thermal based actuation</td></tr>
</table>

The above schemes operate through a common *master relay* employed for a bank of transformers. All the protective schemes mentioned initiate this master relay which provides trip juice (DC) for all breakers and trips the supply for the whole substation.

The operational engineer should isolate the faulty transformer and bring back other healthy transformers and put the system network into service.

If only the winding temperature relay acts and initiates the master relay with no action from other relays, then it should have been only due to overload; further, this action occurred without giving consideration to oil temperature alarm received earlier. In this case the transformer can be put back into service by duly limiting the loads. Before putting the faulty transformer back into service, all health-related tests should be conducted as described in Section 2.2 and it should be confirmed as okay.

5

Inferences from Observed Phenomena and Some Case Studies

This section covers some of the inferences drawn from magnetic balance tests, observations made on voltages during Horn Gap (HG) fuse (HV-side fuse) failures, on sheared-delta connections and inferences drawn from the two-phasing phenomenon. Some case studies on power and distribution transformers are also described. Measures for performance enhancement by design to accommodate the field conditions of distribution transformers are also discussed.

5.1 Inferences from the Transformer Phenomena

5.1.1 Magnetic Balance Test

The magnetic balance test is conducted on transformers to identify inter-turn faults and magnetic imbalance. This is only an indicative test (observation of the phenomena), for which the following procedures are prescribed along with the precautions to be taken.

1. The voltage applied on the respective side (HV or LV) should be below the % impedance volts.

Caution: The application of a high/rated voltage in LV winding may induce a high voltage in HV winding, which is a hazard to safety while conducting tests.

For example, for	11 kV/400 V distribution transformer		
	HV side voltage to be applied	=	250 volts
	LV side voltage to be applied	=	20 volts

2. The voltage should be applied on one phase coil of the winding and induced voltage observed on the other phases. In case of Δ-connected HV windings, the voltage should be applied between two phases; for Y-connected LV winding, the voltage is to be applied between the phase and the neutral.
3. A series lamp (one or two in number depending upon the applied voltage, say 25 watts) is connected in series with the supply side in case of HV side application to limit the higher current that may flow in case of a short circuit of other windings (Fig. 5.1). The lamp / lamps will glow brightly if the readings in other two phases are zero / very low, indicating damage in the windings of concerned phases.
4. In a healthy transformer, if voltage is applied in the central limb, the voltage measured on the other end phases will be equal and will be about half the applied voltage. If voltage is applied on the end phase, the readings obtained on the middle and the other extreme phase will not be equal but will differ slightly, with higher voltage on the middle phase. In any case, some of the induced voltages must be equal to the applied voltage.
5. High impedance voltmeters like AVO meters may be used for the test.

Magnetic balance test for HV side

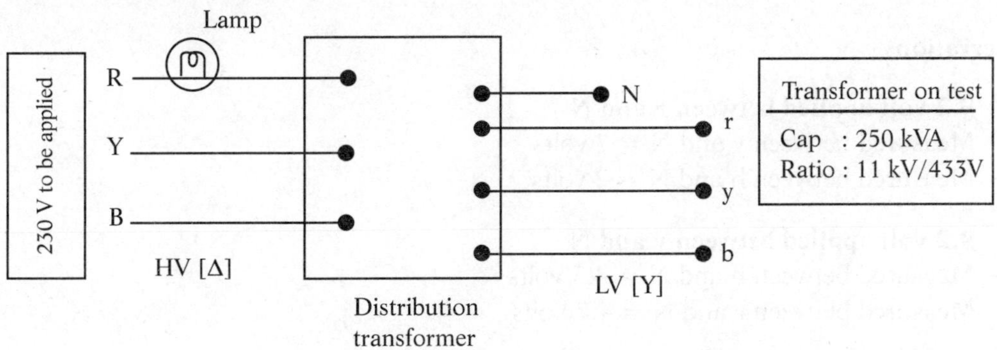

Fig. 5.1 *Magnetic balance test for HV side*

Observations

1. **250 V applied between HV R and Y phases (R phase coil)**

   Measured between Y and B = 190 volts (reading on Y phase coil)

   Measured between B and R = 57 volts (reading on B phase coil)

2. **250 V applied between HV Y and B phases (Y phase coil)**

   Measured between B and R = 120 volts (reading on B phase coil)

   Measured between R and Y = 127 volts (reading on R phase coil)

3. **250 V applied between HV B and R phases (B phase coil)**
   Measured between R and Y = 62 volts (reading on R phase coil)
   Measured between Y and B = 186 volts (reading on Y phase coil)

Magnetic balance test for LV side

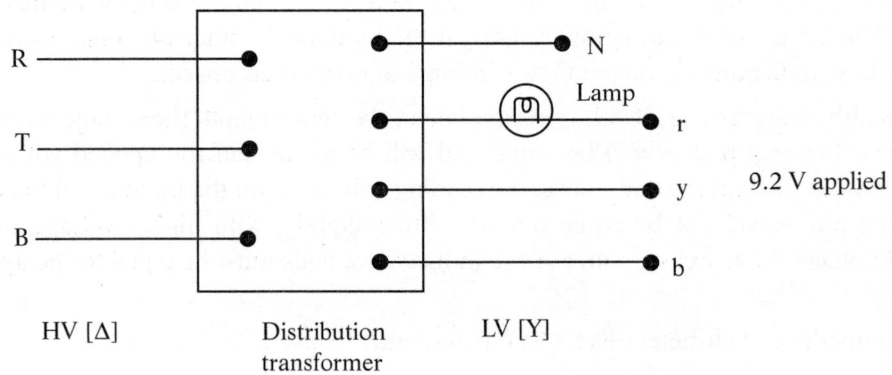

Fig. 5.2 *Magnetic balance test for LV side*

Observations

1. **9.2 volt applied between r and N**
   Measured between y and N = 7 volts
   Measured between b and N = 2 volts

2. **9.2 volt applied between y and N**
   Measured between b and N = 4.3 volts
   Measured between r and N = 4.7 volts

3. **9.2 volt applied between b and N**
   Measured between r and N = 2.2 volts
   Measured between y and N = 6.8 volts

5.1.2 Observation on LT Voltage for HV Side Fuse Failure

For application of 11 kV on HV side (Fig. 5.3):

(i) R phase fuse blown out on HV

LV side reading : r - N : 126 volts

y - N : 256 volts

b - N : 131 volts

(ii) Y-phase fuse blown out on HV

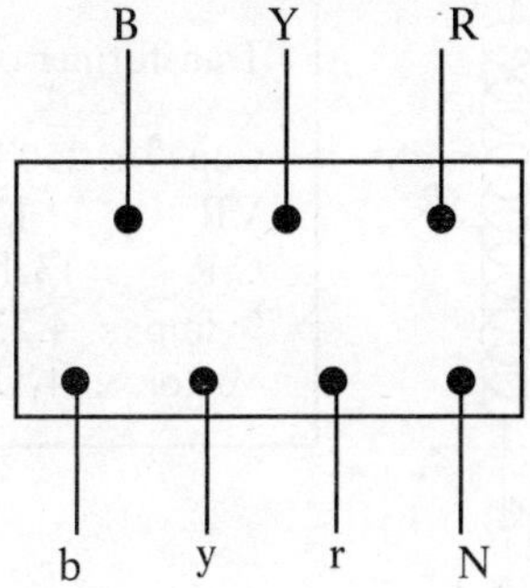

Transformer on test
Cap : 63 kVA V. R : 11 kV / 433V C.R : 3.3 / 84 A

Fig. 5.3

LV side reading : r - N : 93 volts

y - N : 151 volts

b - N : 244 volts

(iii) B-phase fuse blown out on HV

LV side reading : r - N : 254 volts

y - N : 164 volts

b - N : 94 volts

5.1.3 Delta Lead Cut Off: Inference and Observation on LT Side

Refer to Fig. 5.4 for observation on LT side for one phase.

V_{Line} applied = 418 volts.

(i) Observations when R phase delta at location 1 is cut off:

(a) LT reading on open circuit	(b) LT reading on short circuit
r – N : 9.5 volts y – N : 9.5 volts b – N : 9.5 volts r – y : 16.6 volts y – b : 16.6 volts b – r : 16.6 volts	r Phase : 5 A y Phase : 360 A b Phase : 360 A Neutral : 360 A

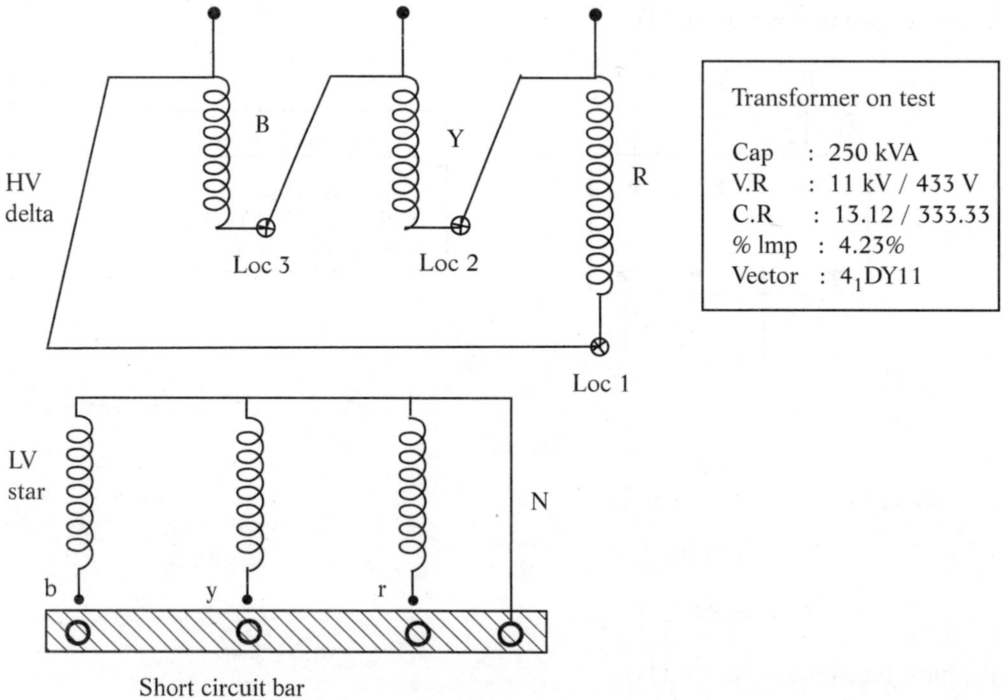

Fig. 5.4

(ii) Observation when Y-phase delta at location 2 is cut off:

(a) Open circuit reading LT	(b) Short circuit reading LT
r – N : 9.7 volts y – N : 9.7 volts b – N : 9.7 volts r – y : 16.8 volts y – b : 16.8 volts b – r : 16.8 volts	r phase : 360 A y phase : 5 A b phase : 360 A Neutral : 360 A

(iii) Observation when B-phase delta at location 3 is cut off:

(a) Open circuit reading LT	(b) Short circuit reading LT
r – N : 9.7 volts y – N : 9.7 volts b – N : 9.7 volts r – y : 16.9 volts y – b : 16.9 volts b – r : 16.9 volts	r phase : 360 A y phase : 360 A b phase : 5 A Neutral : 360 A

Refer to Fig. 5.4 for observation when delta lead is cut off in two locations, 1 and 2 (3-phase 418 V applied on HT):

HV side	:	Only R-phase coil is energised as connected between R and B phase supply (like a single phase transformer). Therefore, the voltage applied to coil is
R phase coil	:	418 volts [R and B]
Y phase coil	:	0 volts
B phase coil	:	0 volts

(a) Observation of open circuit reading on LT side:

r phase coil	:	9.5 volts [r – N]
y phase coil	:	0 volts [y – N]
b phase coil	:	0 volts [b – N]

(b) Short circuit test

R phase = 360 A

N = 360 A

Vector diagram:

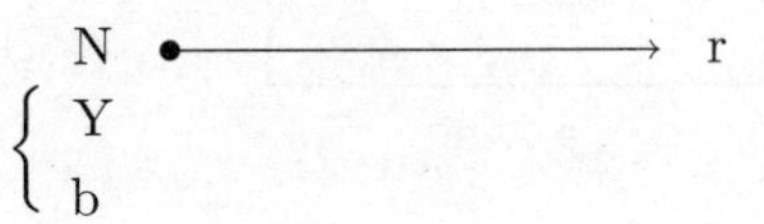

5.1.4 Two-phasing Inference

In some of the utilities, two-phasing is adopted as described below (Fig. 5.5 & 5.6) to contain loads during peak hours and to maintain frequency. The inference of such practice is described as follows.

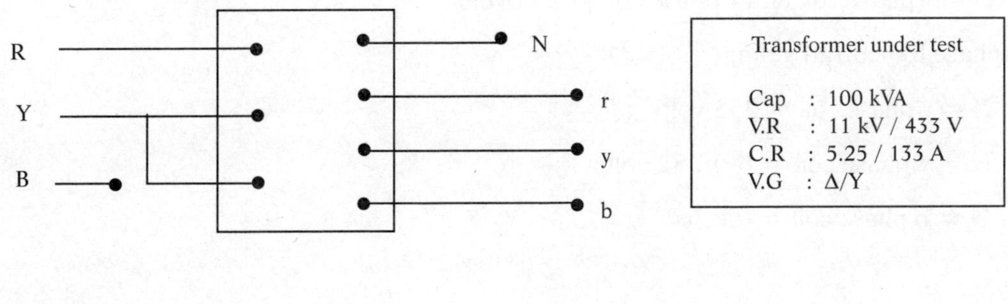

Fig. 5.5 *Two-phasing scheme*

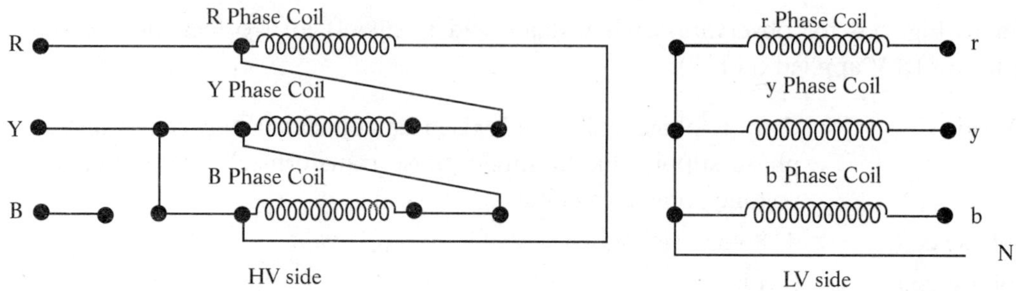

Fig. 5.6 *Physical application of windings*

Observation

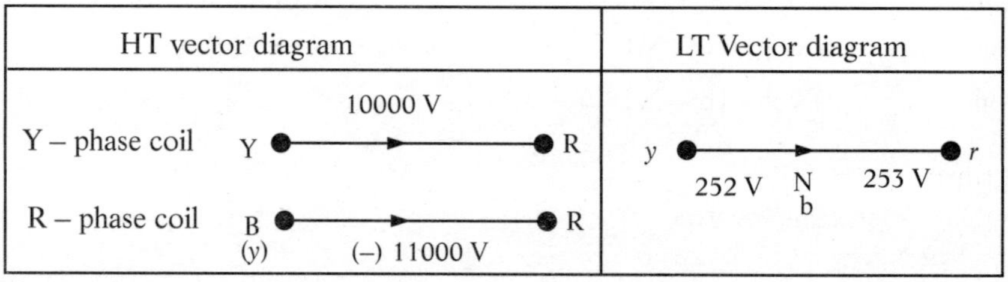

H.T :

R–Y = Y phase coil = 11000 volts

Y–B = B phase coil = 0 volts

B–R = R phase coil = (–) 11000 volts

LT (phase to phase) voltage :

r–y = r phase coil and y phase coil = 505 volts

y–b = y phase coil and b phase coil = (–)252 volts

b–r = b-phase coil and r phase coil = 253 volts

LT (phase to neutral) voltage :

r–N = r phase coil = 253 volts

y–N = y phase coil = (–) 252 volts

b–N = b phase coil = 0 volts

5.2 Case Studies on Power Transformer

Case 1: A message was received that the Buchholz relay top and bottom had acted in a 66 / 11 kV, 5 MVA power transformer, resulting in the tripping of a sub-station through master relay. The necessary solutions were prepared and the presence of gases was tested. On reaching the spot, it was found that the Buchholz relay window was quite clear and no gas was found on opening the test cock. Examination of the conservator revealed that no oil was found on the conservator and the magnetic oil gauge (MOG) showed full oil level. Hence, it was inferred that there were no faults in the transformer. The oil was topped up and the power transformer was put back into service.

The incident is a case of misleading indication of MOG. The draining of oil for frequent testing had resulted in the depletion of oil level, but the misleading indication of full level by MOG prompted the maintenance staff to drain the oil frequently for sampling without topping it up.

Hence, the maintenance practices as referred to in Section 3.1.6 and in Section 3.3.1 should be adhered to. Another inference is that 'low oil level' alarm was not working due to a defect in MOG; the Buchholz relay hence acted as a backup protection to indicate depletion of oil level.

Case 2 A power transformer of 5 MVA 66 / 11 kV failed and on examination, the magnetic circuit was found to have failed with some of the core bolts charred. This would be due to unequal magnetic forces causing local heating because of Eddy currents. The practice of two-phase feeding was being adopted on the 11 kV feeders during peak hours to contain loads, which had caused the unbalance in the magnetic circuit. This type of inference has been described in Section 5.1.

Case 3 In a 22 / 11 kV 1 MVA power transformer, the wrong identification of neutral bushing resulted in a fault. The conventional practice is to have the neutral on the extreme right-hand side, facing the LV side of the transformer. This transformer, however, was of a foreign-make, and when it was transported from the generation station and duly overhauled, it was found that the neutral bushing was kept on the left-hand side with arrangement of phase bushings as shown in Fig. 5.7.

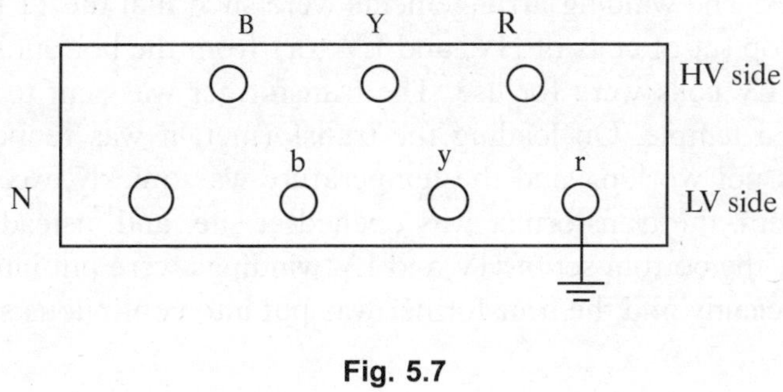

Fig. 5.7

Though the bushing was marked as N, it was completely erased during painting before commissioning, and the earthing was done on the right end of the bushing ('r' phase bushing) due to conventional procedure. This resulted in the short circuiting of the r-phase while charging. Luckily, the transformer escaped without failure due to quick action of the protection schemes and tripping of the sending end feeder breaker.

Case 4 A new 100 MVA 230 / 110 kV auto transformer was supplied and commissioned. All pre-commissioning tests went well, but while energising the auto, the master relay acted up through the differential relay. Again, all the tests were conducted but no fault was detected. A decision was taken by the protection and operation engineers to charge the auto again. The auto stood all right, but it withstood only a month of service before failing by the action of the differential relay. The manufacturing company examined the units and found inter-turn short on R-phase winding for 5 turns only.

The inference from this case is that the weakness in insulation got aggravated in the course of one month of service, and that the differential relay being very sensitive, resulted in the identification of the fault.

Case 5 A 33 / 110 kV substation has two 3 MVA power transformers feeding four 11 kV feeders. Back feeding was extended from another substation from a cut point located at 500 metres away. However, the back feeding phase identity did not match that of the bus. The feeder and its breaker were kept open because of the back feeding. During normalisation after shut down one day, the operator engaged in normalising all 11 kV feeders inadvertently normalised this feeder as well. Because of the different phase identity, there were 'dead shorts' on the breaker itself, resulting in the bursting of the breaker and the simultaneous failure of two power transformers. The heavy short circuit forces resulted in the failure of windings and it shattered the coils.

The inference from this is that operational inefficiency resulted in the mishap, and that the windings of the transformers had become too weak to bear the brunt of the short circuits.

Case 6 A 500 kVA 400 V / 33 kV step-up transformer was received from a thermal generating station for use as a step-down transformer of 11 kV / 400 V, de-rating the capacity to 250 kVA. The winding arrangements were such that the 11 kV / 400 V was obtained from the top set of coils of HV, and LV (or) from the bottom set of coils. The top set of HV and LV coils were for use. The transformer was sent to the field for an important event in a temple. On loading the transformer, it was found that the 'ON' cooling system was not working and the temperature was quickly exceeding the limit. Realising the problem, the transformer was opened at site, and instead of utilising the top set of windings, the bottom set of HV and LV windings were put into use. Now, the cooling worked efficiently and the transformer was put into continuous service.

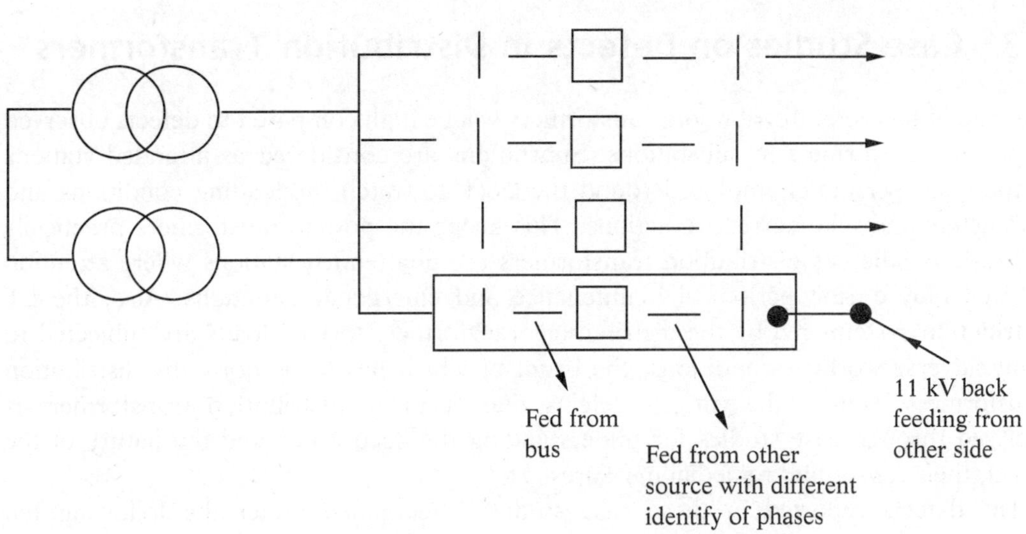

Fig. 5.8

The inference is that the heat absorption was only on the top portion of the transformer with the bottom remaining cool. Hence, 'ON' circulation was not activated. By changing to the bottom set of coils, heat absorption was achieved in the bottom, enabling hot oil movement to the top and its due entry to the cooling radiators, bringing about active ON circulation.

Case 7 In a new substation at an elevated location, a 10 MVA 110 / 11 kV power transformer was installed and put into service. During the working of the substation, one of the operators observed that the surge diverter in the R phase HV was not conducting surges since no readings were observed on its surge counter, whereas surge counters on the other two phases were indicating surge discharges. In the meantime, after one month's service a heavy lightning struck nearby, causing failure of the power transformer. On examination it was found by the company that the R phase lead from the HV bushing to the coil had melted, along with two turns on the edge of the R phase coil winding. It was attended to by the company, and the power transformer was put back into service.

The inference drawn from this is that the surge diverter in R phase did not conduct surges and hence the surge wave created by the lightning strike at a nearby location entered the transformer and damaged its leads and a winding. Hence, it is essential to monitor the surge counters on substation for conduction of surges and record the surge counter readings regularly.

5.3 Case Studies on Defects in Distribution Transformers

The rate of failure of distribution transformers will be high compared to defects observed in power transformers in substations. Substations are considered as attended stations as there are personnel employed round the clock to watch the loading conditions and parameters and take corrective actions. This keeps the power transformers practically failure-free, whereas distribution transformers are unattended stations where attention is given only during periodical maintenance and emergency situations. Also, the LT distribution systems fed by the distribution transformers to feed loads are subjected to many adverse working conditions, the brunt of which has to be borne by distribution transformers. Hence, the various defects that occur in distribution transformers is analysed through case studies for understanding the occurrence and the nature of the defect, their causes and remedial measures.

The defects reported in these case studies is classified under the following ten categories.

1. Manufacturing defects
2. Improper handling of bushing connections
3. Defect due to ageing
4. Defect in HT distribution system
5. Defects arising after repairs
6. Defects in oil handling systems
7. Defects in LT system
8. Improper earthing system
9. Defects due to surges and lightning
10. Defects due to air trap

A case of excellence in performance of a transformer

Before proceeding with the analysis of defects through case studies, it is worthwhile to mention the case of manufacturing excellence of a transformer that withstood like a rock a load twice its capacity for three days to cater to the needs of a festive occasion of an area.

The transformer had a capacity of 50 kVA. It was decided to meet the eventuality of overloading with this transformer itself by providing adequate cooling arrangements. Additional cooling facility was provided by sprinkling water on the surface and through pedestal fans. Higher fuse rating was adopted. Personnel were posted round the clock to monitor the arrangements and loading. The loading was recorded through tong test reading at frequent intervals which went up to 130 A, and this loading was continued for three days without any abnormal occurrence. The maximum temperature recorded was only 80°C.

On the completion of the festivals requirement, the transformer was examined to discover how it withstood the excess demand, and the following was observed.

1. Factor of safety adopted for the size of the conductor and insulation was around 2.5.
2. The cooling system provided was excellent with large space around the core in the container tank, with large quantity of oil.
3. Height of the top cover was more than half the height of core, providing enough space for ON circulation.

Case 1: Manufacturing defects - Example 1

A 250 kVA 11kV/433V transformer in its working condition was reported to have all its three HG fuses blown out, occurring at an interval of 2 hours during continuous full load operation on the occasion of a local festival.

The transformer was examined at the site, since replacement was not possible. Casual observation revealed that the full load oil level on the conservator was below the height of HV bushings. This led to the non availability of oil inside the bushings for a height of 5 cm from the top, resulting in heating of the leads in the uncovered portion. Transmission of this heat to the jumpers resulted in the blow out of HG fuses in a specific time period. As a temporary measure, in view of the festival season, oil was topped up to 90% of the conservator tank, and its breather removed and closed with a wire mesh. This ensured sufficient oil inside the bushings, and there was no blowing out of the fuses thereafter.

The transformer was later examined and it was found that the insulation sleeves inside the bushings were charred to a length of 5 cm from the top. New sleeves were provided. Arrangements were made to raise the height of conservator.

Case 2: Manufacturing defects - Example 2

A particular consignment of 50 kVA 11kV/433V transformers failed frequently and unusual heat was observed during the working condition of the transformers at locations where they were subjected to full capacity loading. On examination at the laboratory, it was found that the oil in the tank with core-coil unit inside was only 80 litres as against the 200 litres mentioned in the name plate. Inadequacy of oil content deprived the normal rate of cooling. Sustained heat caused insulation failure.

The supplying company then provided a higher size tank for the core coil unit, with sufficient quantity of oil for the ten transformers supplied under the consignment.

Case 3: Improper handling of bushing connections - Example 1

It was reported that the A-phase LT bushing rod was worn out and the voltage of the A-phase dropped down considerably when the phase was loaded. The yellow tape wrapped around the rod was completely charred and had become brittle. The unit was examined at the site. It was observed that the coil connection from A-phase LT coil with bushing had become loose inside. The coils were tested and found to be okay. The loosening had occurred due to improper handling of external jumper connections.

The LT A-phase rod was replaced and it was ensured that all other connections to the bushings inside were made and checked. Special additional nuts were provided externally so that the fixing nut and its check nut were not disturbed. It was recommended that jumper connections be made in between the additional two nuts without disturbing the fixing and check nut. Further, it was suggested that two spanners be used to handle the connections.

Case 4: Improper handling of bushing connections - Example 2

It was reported that in a 100kVA 22kV/400V transformer, the no-load voltage on all LT phases was all right, but it dropped down when load was connected. Appreciable current was measured in the neutral and heat was also observed in the neutral bushing rod. The transformer was examined at site. The earth resistance was measured and found to be a satisfactory 3.5 ohms. It was observed that the neutral connection from the LT winding to the neutral bushing was loose. This had occurred while removing the jumper connections of the neutral bushing for measuring the IR value of the transformer during maintenance.

Proper tightening of connections was made inside, and all abnormalities were set right. Additional nuts were provided for neutral bushing and for other phase bushings as well, and the personnel in charge were advised to handle the unit carefully thereafter.

Case 5: Improper handling of bushing connections - Example 3

In the case of a 63kVA 11kV/433V transformer, it was reported that an unequal voltage was observed on the LT system as if the HG fuse had blown out in one of the phases. While meggering, no continuity was observed on the HT B-phase with other phases. The transformer was examined at site. On opening the top cover, it was found that the HT B phase coil connection to the bushing rod from the hole of the rod to which the lead was kept soldered was cut, causing an open circuit.

It was inferred that during annual maintenance, while checking the jumper connections on HV side, improper handling of the nuts at the top of the bushing had resulted in the internal connection of B-phase giving way from the soldered point. The lead connections inside were also taut with inadequate flexibility, which caused the occurrence.

The internal connections were made flexible by increasing their length and they were properly soldered with the bushing rod. The other two connections were also checked. Directions were given to handle the bushing connections carefully. Improper handling could also damage the bushing due to the breaking of the groves at its top, provided to let in the bushing rod.

Case 6: Defect due to ageing

An old 100 kVA 22kV/400V transformer had served for more than 30 years. It was reported that unequal voltages were developing in the transformer; further, the continuity check was found to be okay.

On examining the transformer at site, it was found that the delta lead emanating at the bottom of the C-phase coil to connect to the top of the A-phase was found cut at its soldered point near the C-phase coil. It was also observed that all the joints were brittle. Heavy sludges due to ageing were found adhering to the coils.

The cut delta lead was properly soldered and rectified temporarily. It was later brought at leisure to the laboratory, where a complete overhauling was carried out. All sludges were removed; coil was varnished with insulation; heat treatment was given for drying out; and new tested oil was topped up and sent to service after hot oil circulation and testing. The transformer so overhauled and maintained would definitely be in service for another 20 years or more. The continuity was recorded as okay since the cut was on the delta point only.

It is advisable to check formation of sludges and condition of oil by sample obtained from drain valve for aged transformers.

Case 7: Defects in HT distribution

A sole distribution transformer of capacity 63 kVA, 11kV/433V was fed by an HT line laid for a length of 2 kilometres. There were no other transformers connected on this line. Unequal voltage was reported on this transformer, but on performing continuity test, it was found to be okay. This transformer was replaced with another one. But while energising the new transformer, the same defect was observed again.

The first transformer was tested in the lab and found to be okay. This was informed to the field team and they were asked to patrol the HT line properly. It was then discovered that an outgoing jumper of the Y-phase was cut off in the midst of a lake and the conductor was lying exposed in the rain water. The controlling breaker has not tripped for this condition. The lake being in the midst of the line route, prevented intensive patrolling, leading to the defect going unnoticed.

This case is not due to fault on the transformer but due to HT line defect. It may be noted that a cut off in jumper connections in the HT line will also lead to such a misleading observation.

Case 8: Defects after repairs – Example 1

A 250 kVA, 11 kV/433V transformer was repaired and sent to distribution for use. On erection and commissioning, no continuity was observed on HV winding and hence was brought back. It was found on examination that the delta leads soldering joint at A and B phases were sheared during transport, causing the discontinuity. Improper soldering had caused the shearing during transportation.

The sheared delta connections was properly soldered and inserted inside tube sleeves and fastened with supports to maintain rigidity during transportation.

Case 9: Defects after repairs – Example 2

One 50 kVA 11 kV/400 V transformer was sent after repairs for use. A month after, unusual heat was reported on the main tank but the radiators were found to be cool.

On opening the top cover of the transformer, it was noticed that a handful of cotton waste was blocking the entry point of the oil into the radiator at the top header. This had blocked the cooling system, causing the observed phenomenon. Since the system was a single entry and retrieval system, the cotton waste used during repairs that had fallen inside had gone unnoticed. Due to circulation current it had moved up and caused blockage of the oil entry.

The blocking was removed and the radiator was checked and found to be okay. Care should be exercised while carrying out repairs. All tools used should be tied by hand with cotton tape to avoid such mishaps.

Case 10: Defects in oil handling system – Example 1

A 50 kVA, 22kV/433V transformer was brought to the lab due to blowing out of the HG fuse in the C-phase. The IR values and continuity were found to be okay.

On examination, the 2nd and 4th stack coils on the HV side C-phase were found to have failed due to inter-turn short. The oil sample was not satisfactory. The crackle test showed that heavy moisture was present. The breather arrangement was bad and the diaphragm of the explosion vent was found broken, which had led to free entry of moisture. This in turn led to the failure of inter-turn and inter-layer insulation. Failure did not involve earth and hence the megger test was found okay.

The failed stacks were replaced and the whole core coil unit was dried out and circulated with hot oil, tested and sent to use after revamping the breather and the vent pipe.

It is essential that all the components of the oil handling system be maintained properly.

Case 11: Defects in oil handling system – Example 2

While carrying out a periodical maintenance in a distribution transformer, the air release plug was opened but no oil came out. The top lid of the conservator was opened and the oil level was checked with dip stick. The stick showed only residual oil. The oil level in the gauge glass (plain type) showed full level. On completely opening the air release plug, oil was found up to the core level only.

Oil samples were taken frequently for test and there was leakage in the side-mounted LT bushings. This leakage had stopped subsequently. The misleading oil indication through gauge glass resulted in this slippage. Further depletion of oil would have caused failure of the windings.

The oil was topped up to the required level and the gauge glass was attended to by clearing the blocked holes inside and by cleaning the gauge glass piece.

It is essential that oil level be checked by visual examination in the conservator. This sort of occurrence was reported in another power transformer where the depletion of oil level had caused the Buchholz relay floats to operate, as the oil had come down the relay and the buoyancy of the float had been lost. This had also occurred due to misleading oil level indication, with the MOG showing level as 'full'.

Case 12: Defective LT system – Example 1

It was reported that all the HG fuses were blown out in a 250 kVA, 11kV/433V transformer. The IR value on the HT side was okay, but the value on the LT side was zero. The non availability of transformer for replacement led to an examination at site.

On finding the above observations correct, as a trial measure, all the LT jumper connections including the neutral were removed. The IR values on the LT side was measured on the bushings and was found to be okay. But when measured on the removed jumper leads, it was found to be zero. The transformer was charged idly without the jumpers, and was found to be okay. The voltages on the LT side were also measured and found as okay. The problem was only on the LT leads and not on the transformer. The LT cable leads were replaced and distribution put into service.

The present-day arrangement of the LT bus bar system makes such exigencies (with the mounting insulators becoming defective) uncommon.

Case 13: Defective LT system – Example 2

A transformer of 100kVA, 11kV/400V was received at the lab with its HG fuse blown out in A-phase. The HT to earth IR value was zero. Oil was spurting out through the vent pipe. This was the second occurrence of such an incidence within five months.

The unit was examined to find a complete shattering of HV A-phase coil and contact of earth with the shattered winding. Heavy shattering had resulted in the oil spurting out. While verifying the previous occurrences of transformer failure, it was found that a similar failure had occurred in the same phase earlier.

A patrol of the LT system of the distribution was initiated. It was found that the neutral wire had run to the bottom of the insulator pin resulting in low clearance. At a particular location, the sag was too much. The fuse provided was of a 'current carrying wire' and not fuse wire. This had resulted in the following: frequent contact of the A-phase conductor with neutral and separation; fuse not blowing out; frequent stresses due to short circuit on A-phase winding; weakened winding, and finally failure. The LT system which was defective was revamped, and such failure did not occur thereafter.

Case 14: Defective LT system – Example 3

A transformer of 63kVA, 22kV/433V was brought to the lab with its C-phase HG fuse blown out. But the HT-to-earth IR value was found to be okay. On examination, it was found that inter-turn short circuit had occurred on the 3rd stack of the 6-stack C-phase winding. The insulation of all other stacks were found charred due to heat and were brittle.

All the six stacks of the C-phase winding were replaced and the transformer put back to use. The manner of failure indicated sustained overloading of the winding. The analysis carried out by the field engineers indicated that there was continuous overloading in the C-phase, as observed by tong test reading. The peak hour reading measured was: at A-phase 60 A, B-phase 20 A and C-phase 100 A, against a per phase permitted maximum loading of 84 A. Because of the two-phasing operation being resorted to during peak hours (night 6 PM to 10 PM) in order to provide single phase supply for the lighting loads (refer the inference at the end of this section), the single phase service connections were dumped in C-phase, resulting in overload. Remedial measure by rearranging the loading in the phases were resorted to. It may be noted that in this case the total loading was within the capacity of the transformer.

Case 15: Improper earthing system

A 100 kVA, 11kV/433V transformer failed, with the C-phase HG fuse blowing out and showing IR value to earth as zero. An examination revealed that the top coil of the C-phase HV winding and its leads and also the top portion of LT coil insulation along with the laminations were charred.

This is a rare phenomenon which occurred due to rise of earth potential and consequent partial discharges near the C-phase coil.

The site was examined and the earth resistance was found to be 100 ohms with all the earth pipes completely cut off. The earth system was completely revamped by replacing all the earth pipes and earthing conductors to avoid recurrence of this type of failure.

Case 16: Defects due to surges and lightning

One 100 kVA 11 kV/433V transformer in a hilly area was found failed with all the three HG fuses blown out. The IR value of HT to earth was find to be zero. Oil had spurted out through the vent pipe. A similar happening had occurred at the same location a month ago.

On examination of the failure, it was found that the B-phase HV side top stack coil and bottom stack coil and the C-phase HT top stack coil were heavily shattered. Adjacent stacks also were affected. The colour of the oil was dark brown.

This sort of failure is to be attributed to the entry of surges (high frequency, high voltage waves) into the transformer through the HT line. In this case, the surge had entered through the B- and C-phase lines, resulting in the shattering of the end coils.

A similar type of damage had occurred in the previous failure at this location. The area was found to be prone to heavy lightning. On examining the location, no lightning arrestor was found in the C-phase. The other two lightning arrestors were also not satisfactory. The earthing systems of LAs required revamping.

Hence, all lightning arrestors (LAs) on the HV side and LV side were replaced with new ones and the earth wires from LAs to earth pipe were renewed. There were no such occurrences reported thereafter.

A direct stroke of lightning may hit the vicinity of the transformer or the transformer structure itself. Hence, at such locations it is better to erect an earth mast at a distance of one or two poles from the transformer and connect it properly to the earth with a suitable earthing system of very low earth resistance value.

Case 17: Defects due to air trap - Example 1

It was reported that the HG fuse of the transformer in A-phase was blowing out every five hours. The transformer was continuously feeding the loads without any problem and the test values were okay.

An examination was carried out at site. The transformer was found to have all the bushings mounted vertically in its top cover. The top cover with the core was raised form the tank and were no defects. The HT bushing rod was dismantled on the A-phase to verify any partial damage of the connection and it was found to be okay. But the insulation sleeve was charred to a considerable length from the connected bushing rod.

This clearly indicated that air had entered and was trapped in the A-phase bushing, pushing the oil down. This had resulted in undue heating of the leads in that phase, and accumulation of this heat in the HG fuse had resulted in the blowing out of the A-phase fuse in specifics duration of time.

The lead insulation was changed. The maintenance staff was advised to release air through the air release plug during periodical maintenance work.

Case 18: Defects due to air trap - Example 2

This case is an almost unbelievably peculiar phenomenon that occurred in a single-phase 15 kVA small transformer located near the seashore. The saline content of the humidity had corroded all the screwable threaded fittings.

It was reported that oil sample could not be taken from the bottom sampling valve of the transformer. The gauge glass indicator was completely flooded with oil above the full load level mark. The conservator top lid could not be opened due to corrosion.

While attending the reported defect, the above observations were found to be completely true. With much difficulty the corroded top cover lid was slowly opened, only to be thrown out with a great force along with a loud sound, landing about 30 meters away. The gauge glass level came down to normal, and oil could then flow out from the bottom sampling valve.

This phenomenon is explained with the following reasoning:

A column of air had got trapped at the bottom of the transformer, pushing the oil up to a height more than the height of the valve. The air refused to be released through the side valve. But when the top lid was opened, the large quantity of trapped air rushed out through the top lid.

Opening of the top lid of conservator is also a good air-releasing exercise and should be undertaken to avoid air trap. This will also enable the checking of the oil level.

5.4 Conclusion Drawn from Case Studies

1. Proper maintenance of the oil-handling system is a must for transformers.
2. Proper construction and maintenance of LT system helps in avoiding the failure of the transformer.
3. Keeping proper loadings within the capacity of the transformer and within temperature limits will enhance transformer life.
4. Earthing systems and lightning arrestors, if maintained properly, will enhance the performance of transformers.
5. It may also be noticed that in case of failures, the megger values of IR and continuity will not reveal the condition of failure; even attributes will show up as if the transformer was good. This can be seen in some of the case studies.

5.5 Performance Enhancement by Design of Distribution Transformers for Field Working Condition

The discussion to follow will elaborate the performance enhancement of distribution transformers in the prevailing working conditions of the LT distribution system by adopting some design standards during manufacturing.

Working conditions

The distribution transformers erected and maintained by the utilities are unattended stations with no possible control over the working parameters, and hence they bear the brunt of the working conditions. Distribution transformers maintained by the HT industries, on the other hand, are well attended and well regulated for meeting the loads of industries.

The adverse aspects of the prevailing working conditions of distribution transformers maintained by the power supply utilities supplying low tension loads are listed below:

- Load conditions deviating from the maximum efficiency area of the transformer
- Actual loading being greater than contracted loads
- Unequal loading on the three phases, resulting in overloading in one or more phases
- Possible short circuit between phase conductors (or) between phase and neutral/earth
- Impact of external and internal surges

- Absorption of moisture and impurities
- Increase of acidity due to oxidation process enabled by the heat in the oil along with oxygen in the air

Functional accessories of distribution transformers

The functional accessories of a distribution transformer are:

- Magnetic core, designed based on a working flux density
- HT and LT coils with their number of turns for the operating voltages and with the size of conductors based on the current handled according to their capacity rating
- Oil used as an insulant and coolant

The other supporting accessories are :

- Main tank where functional accessories are placed
- Oil handling systems conservator, breather, oil level indicator
- Protective systems, namely explosion vent and air release plug

Design based on maximum efficiency

It is not essential that the maximum efficiency of a distribution transformer occur at full load. The resistive part of the winding is responsible for the copper loss, which is proportional to the load and working power factor. The losses occurring due to hysteresis and eddy current (iron) loss is fixed.

The efficiency is maximum when iron loss = copper loss.

$$\text{Efficiency} = \left[\frac{1 - \text{Losses}}{\text{Input}}\right] \times 100$$

where

$$\text{Losses} = \text{Fixed Loss} + \text{Copper Loss} \times \% \text{ of Load.}$$

$$\text{Input} = \text{Rated kVA} \times \text{P.F.} \times \% \text{ of Load.}$$

The (Losses ÷ Input) can be kept less than 0.02. Hence the fixed losses can be kept to the minimum by designing the magnetic circuit of, say x watts. The copper losses can be kept equal to iron loss between 90% to 100% of load by designing proper conductor sizes. Let the number of turns and mean length be say, y watts. The input power computation for the above design can be done at 0.8 PF lag, say P watts, so that

$$\frac{x+y}{P} \leq 0.02.$$

Thus maximum efficiency can be maintained around 90% of load in a varying power factor range of 0.8 to unity.

Design based on load condition

The distribution transformer is generally occasionally subjected to overload condition during peak load hours of the day. Due to financial considerations and competitive market conditions, the factor of safety adopted is generally 1, or even slightly lesser. Overloading is a major factor in distribution transformers not lasting up to their expected design life of 20–25 years.

Hence a factor of safety of 1.2 is suggested in the following design.

1. Size of conductors of winding sufficiently big to meet the overload conditions
2. Insulating materials of required quality to meet thermal stresses
3. Higher quantity of oil and volume of main tank, with sufficient spacing around the core and with large enough height from the core of top cover, preferably half the height of the core
4. Improved cooling system to dissipate heat with double entry and retrieval system

Meeting short circuit forces

The occurrence of a fault in low tension line results in a heavy flow of current, of the order of 10-15 times the load current. The protective arrangement clears the fault. The windings are expected to withstand the heavy forces during the period between the occurrence of the fault and its isolation. The mechanical rigidity of the winding should be made strong with spacers and vertical duct arrangement, properly introduced with rigid fixing of the windings between the top and bottom frames. Also, by considerably increasing the percentage impedance of the winding in design, short circuit stresses can be reduced by limiting the short circuit current. This is explained below for a 100 kVA 11kV/400V, 5.25/133 A distribution transformer.

A. Level of short circuit for a for 4% impedance

Per unit impedance	=	0.04
Per unit short circuit current I_{sc}	=	$1.0 \div 0.04 = 25$ p.u.
Base current p.u.	=	5.25 A (HT side)
Therefore, short circuit current	=	I_{sc} p.u. $\times$ base current
	=	25×5.25
	=	140 A on HT side.
Short circuit force	=	αI_{sc}^2
	=	$\alpha\ (140)^2 = 19600$ units.

B. Level of short circuit force for 5% impedance

Per unit $I_{sc} = 1.0 \div 0.05 = 20.$

Short circuit current = 20 × 5.25 = 112.5 amps on HT side.

Short circuit force = $(112.5)^2$ i.e., 12656 units .

Increasing the percentage impedance from 4% to 5% will short circuit forces by 35%

$$\text{i.e., by } \left(\frac{19600 - 12656}{19600} \times 100\right).$$

The design may be carried out by choosing proper requirement of turns and mean length. However, this will increase power loss in the transformer, which should also be kept in limits to achieve the desired efficiency.

Consideration of surges

The lightning arrestors employed independently for the three phases, before and after the distribution transformer, diverts the high-voltage, high frequency surges coming on the line to the earth and saves the windings from damage. Proper selection of lightning arrestors to prevent the damage of the winding is one external solution, the strengthening of winding is another. Special end windings with lesser number of turns but with more insulation are placed on the top and bottom ends of the winding with special mechanical rigidity. Though this is in practice for power transformers in sub-stations, it can be adopted for distribution transformers also and the basic impulse insulation level (BIL) value of the transformer can be made sufficiently high and duly coordinated with the lightning arrestors.

Modification in oil handling system

Absorption of impurities of moisture and increase of acidity in oil are the other important factors for the non-performance of distribution transformers. It is important that the working condition of the transformer be linked with atmospheric pressure to avoid damages due to undue pressures. Hence, the conservator top portion inside is maintained at atmospheric pressure. Any improper maintenance of the breather unit with silica gel results in impurities, with and moisture entering the transformer and getting absorbed in the windings. Hence, the following is suggested.

A. Use of bellows in the conservator

1. The bellow or bladder, made up of material non-reactive to oil, may be placed inside the conservator and the breathing system connected to this bellow only. This system will maintain the volume of oil in expansion and contraction without direct contact with air.

2. A side cover for the conservator must be available to open and check the condition of the bellows/bladder and to clean the conservator.
3. Since any direct contact of air/oxygen with hot oil is avoided, increase of acidity due to oxidation is controlled.

B. Modification of explosion vent

The explosion vent uses a shattering type diaphragm to spurt oil during failure. Non replacement of this diaphragm (in its broken condition) results in heavy entry of moisture and suspended impurities. Hence, a non-shattering type vent (spring operated) of a small size (miniature PRV) suitable for distribution transformers is the best alternative to the conventional shattering type vents.

Arrangement to record maximum temperature and maximum load

Since the distribution transformer is an unattended station, it is necessary to record the maximum temperature and load current so as to have an idea of the loading condition while resorting to corrective measures. The following arrangement is recommended.

1. Installation of a dial-type thermometer with maximum temperature indicator and resetting arrangements. This arrangement is to be thought of by providing the thermometer stem directly under the top cover with an oil seal; the top oil temperature is measured without using the thermometer pocket since the availability of oil in thermometer pocket cannot be ensured in field.
2. Maximum demand meter showing ampere loading in all the three phases with maximum ampere indicators along with current transformers (CT) installed in separate enclosures, duly protected from rain and sunlight.

Conclusion

The distribution transformer is subject to many adversities. Frequent failures result in heavy expenditure incurred on replacement, transport and repairs besides attracting the ire of consumers. Hence, an increase in expenditure for improved design is justified when compared to the expenditure incurred in meeting failures. A minimum budget may be worked out for the basic requirement and the procurement may be made accordingly with a permissible margin so as to maintain quality. Inspecting the transformer manufacturer process at factories and conducting strip test after their delivery will ensure that the quality guaranteed by the manufacturer is assured and thus prevent failure in the field. Finally, the installation of a plain-type three-phase oil circuit breaker on the LT side with CTs and overload relays can also be resorted to, at least in urban areas where continuous load is a certainty. The time taken to reach and switch on the breaker will be less than that required to replace blown out fuses. Alternatively, the provision of HRC fuses can be resorted to for efficient fusing.

References

1. *J & P Transformer Book*, 10th Edition, Elsevier Ltd.
2. Z. Khudyakov, *Repair of Power Transformers*, translated by S. Kittel, Mir Publishers, Moscow
3. *Transformers*, Bharat Heavy Electricals Ltd., Bhopal, Tata McGraw Hill
4. Code of Technical Instructions of TNEB
5. Seminar Papers CPRI, Bangalore
6. Standards and practices adopted by power system utilities

List of abbreviations

AC	Alternating current
BDV	Breakdown voltage
BIL	Basic impulse insulation level
CRGOS	Cold rolled grain oriented steel
CT	Current transformer
DC	Direct current
DGA	Dissolved gas analysis
E/L	Earth leakage
EHV	Extra high voltage (110 kV and above)
EMF	Electromotive force
HT	High tension (11, 22, 33 kV)
HV	High voltage
I	Current in amperes
IDMT	Inverse definite minimum time
IR	Insulation resistance (mega ohm)
kVA	Kilo volt ampere (power)
LA	Lightning arrester
LF	Load factor
LHW	Left-hand winding
LT	Low tension (230, 400 volts)
LV	Low voltage
MOG	Magnetic oil gauge
MVA	Mega volt ampere (power)
N/C	Normally close in working condition
N/O	Normally open in working condition
O/L	Overload
OLTC	On load tap changer
PF	Power factor
PI	Polarisation index

PPM	Particles per million
PRV	Pressure relief valve
PT	Potential transformer
PU	Per unit
P_Y	Primary side (winding)
R	Resistance in ohm
RHW	Right-hand winding
SWG	Standard wire gauge
S_Y	Secondary side (winding)
VAR	Volt ampere reactive (KVAR, MVAR)
X	Inductive reactance (ohms)
Y	Star
Z	Impedance in ohm
Δ	Delta
η	Efficiency
RmS	Root mean square value
X_C	Capacitive reactance (ohms)
I	Current in amperes
A	Representing current in amperes
kV	Kilo volt
WTI	Winding temperature indicator
KOH	Potassium hydroxide
N	Number of turns in winding
N_P or N_{PY}	Primary number of turns
N_S or N_{SY}	Secondary number of turns

Index